Zoomie Pig Odyssey

By

Dan N. Clark, Ed.D

©2025 Dan N. Clark. All rights reserved.

Published by Parker Publishers. No part of this book may be reproduced, stored, or transmitted in any form or by any means, whether electronic, mechanical, photocopying, recording, or otherwise, without prior written permission from the author, Dan N. Clark, except for brief quotations used in reviews or other permitted uses under copyright law.

For permission inquiries, please contact:

ltcdanclark@gmail.com

.

DEDICATION

To my former US Air Force and Philippine Air Force colleagues, and to all the free peoples for whom we swore life oaths to protect from world communism.

Deo Vindice!

ACKNOWLEDGEMENT

Foremost, I wish to thank God Almighty for watching over me during my odyssey through peace and war during my three-decade military career. I'm fairly certain I turned my Guardian Angel into an alcoholic.

Next, I acknowledge the love of my life and wife, Jen, for tolerating my endless stories about the places I went and the people with whom I served long before she entered my life. "The Jenral" was the missing piece that completed the mosaic of my life. She stood by me and kept our family together during difficult times. I cannot imagine life without her. To quote my once-protégé and friend, Dr. James Smitt, "You damned sure out-kicked your coverage marrying her."

Having put my parents through hell worrying about me and the things I refused to discuss, I regret that they did not live to see this book published.

I wish to recognize the members of the 3rd Security Police Group and the Philippine Air Force with whom I served at Clark Air Base in the Republic of the Philippines during 1981–82. Many former colleagues have passed on, including Colonel Gary G. Allison and

Chief Master Sergeant Harley Fields. Despite my not returning as a Security Police officer, Colonel Allison mentored me during my quarter-century as a US Army Military Police officer. Many noncommissioned officers, such as Donald Funk, Garther Halbert, Fred Carrender, Lewis Nunnally, Tom Loprete, and Chief Lippi, were strong leaders, great teachers, and caring role models to countless young people far from their hometowns and families.

Yet my deepest gratitude goes to Donald Quesnell, who saw something in me; he taught and mentored me as one would a younger brother. I learned more useful knowledge from Don than from anyone else during my tour there. We both went on to earn doctorates and retired early from our second careers. Don and I remain in close touch to this day.

I wish to recognize the members of the 7th Security Police Squadron with whom I served at Carswell Air Force Base in Fort Worth, Texas, during 1982–84. A number of them have also passed away, including Frankie Howell, Tom Childs, Dave Lynch, and Chief Charles Jett. I've remained in touch with many peers — Ron Rucker, Don Adams, Mark Spinney, Mark Chaires, Todd Swerske, and Bruce Crosby — and greatly enjoy hearing from them despite our advancing ages and fading memories.

I'm eternally grateful to then-Captain Donald W. Neal, Jr., who was my shift commander at Carswell AFB. He tolerated my antics and led our team well. He continually urged me to seek an officer commission until I enrolled in Army ROTC during graduate school. Don later left active duty to become an attorney and served as legal counsel for several State of Texas agencies. Although we never served together during the ensuing Middle Eastern wars, we always compared notes afterward. He went on to retire as a full colonel from the US Air Force Reserve.

Sergeant Clark guarding the Air Force Thunderbirds in a B-52 hangar the night before the Carswell Air Show in April 1984

FOREWORD

I first pecked out this manuscript on a manual typewriter during college at Middle Tennessee State University in 1985, then set it aside for the next 39 years. With time on my hands during a difficult cancer ordeal in 2024, I began transcribing it into an electronic format for my daughters. During phone calls and email exchanges with former colleagues, many reminded me of events and incidents I needed to add.

I changed some names to protect the guilty. I also omitted most of the sexual and violent incidents, as I contend these are implied and thus irrelevant after so many decades. The latter live on in the memories — and, for some, the nightmares — of those who were there. After I completed the transcription and editing in 2025, several former colleagues reviewed the story, and all felt it warranted publication.

Like many of my colleagues, I went on to serve in the military until retirement. Yet, as humans, we have a naive tendency to freeze people in our mind's eye as we knew them at a given point in time. We all changed dramatically as we served in successive assignments and grew older. Barring early death, our respective military service periods were only 20–30% of our lives; for those who served only four years, it was much less. And as the gray invades our temples and our memories fade, it's important we share our stories before they are lost to the sands of time.

As my father warned me before his death in his 90s, the curse of longevity is outliving so many dear friends before we, in turn, "cross over the river." Ultimately, we did the best we could with the training and resources we had at the time. This is my friends' history of a point and place in time as much as it is mine, so I trust I've done everyone right in telling it. I apologize for any aspects I inadvertently got wrong.

My cancer is now in remission and manageable as a chronic condition. Because I want to help others, I am gifting the net proceeds from this book to cancer research.

As such, I hope you find reading it an enjoyable experience.

Table of Contents

"Zoomie Pig Odyssey, 1980-84" 1
Summer and Fall, 1980
"Zoomie Pig Odyssey, 1980-84" 35
   Philippines, 1981-82
"Zoomie Pig Odyssey, 1980-84" 296
   Carswell AFB, 1982-84
About the Author 408

"Zoomie Pig Odyssey, 1980-84"

Summer and Fall, 1980

Following my graduation from Lawrence County High School in Tennessee, I was torn between the Scylla of having multiple good "what next" options and the Charybdis of not preferring any of them. My senior track season was the highlight of high school and my friends were heading in every direction. Our 274 graduates were roughly split between military enlistments, working locally, and enrolling in college. Still with no decision about college, I settled into working on the farm, fishing, and hanging around with a few friends and my cousins. While I wasn't getting into trouble, partying, or running around with wild characters, I realized I needed to find viable employment soon or I'd have to enroll in college.

I was happy with the idea of farming; however, my parents were not, and they insisted that I attend college. The University of Tennessee-Martin had accepted me and allowed me to walk onto its track or football teams. With improvement, I could earn a scholarship by my second year. With three years of Junior ROTC, I could also enroll in the Army ROTC program as an Early Commissioning Program cadet and commission as an officer in two years. Ultimately, I wanted to major in Wildlife Management and minor in Law

Enforcement. This would qualify me for work as a park ranger, game warden, or in law enforcement.

Unfortunately, my mother was adamant that UT-Martin was too far away and insisted that I enroll at the University of North Alabama, 40 miles away in Florence. She said I would move in with my sister, an absolutely unacceptable option for me; I refused to live around her drunk and pothead friends. I also didn't want to attend UNA since its track team was small and so many other high school classmates were already enrolled or enrolling there. The other option was Middle Tennessee State University, but I wasn't keen on it. My parents finally told me I could not stay home and farm; interest rates were ghastly high, and I lacked collateral to borrow money to buy new farm equipment. The threat of dropping me from their car insurance policy lacked the intended leverage, as I could always switch to a different company. Worse, my grandmother and mother began selling all of my late grandfather's farm equipment, thus eliminating my farming option. They were determined that I attend college and enroll at UNA.

My reluctance to enroll at UNA made no one happy and I felt boxed in. Affording college wasn't the issue since I had around $20,000 saved up to pay for a degree. The issue was my mother's overbearing and increasingly shrill demands to do her bidding with no consideration for what I wanted to do with my life. If I couldn't attend UT-Martin or stay home and farm until I figured things out, my only viable option was to get away for a few years. Enlistment looked like the best road ahead. The week after my 18th birthday, I was plowing corn for a neighbor and heard my late grandfather's voice in my head repeating one word: "Go."

The next Monday, I told my mother I was going to visit friends and drove to Florence. I walked into the Armed Forces Recruiting Station and enlisted in the United States Air Force. With my high Armed Services Vocational Aptitude Battery (ASVAB) score, I wanted to enlist as either a Survival Specialist or a Pararescue Specialist. However, neither specialty was available for at least six months. Since I wanted to ship out to basic training right away, I settled on a career in Security Police or Law Enforcement. I would report to Lackland Air Force Base (AFB), Texas, in September, remain there for SP school, and graduate right before Christmas. With my JROTC completion certificate, I was sworn in as an E-3 (Airman First Class, or A1C). My JROTC certificate also allowed testing to bypass the second half of basic training if I received unanimous approval from my basic training cadre team. The recruiter scheduled my physical for Thursday morning at the Military Entrance and Processing Station (MEPS) in Nashville and provided me with a bus ticket for Wednesday afternoon. My effective enlistment date would be the day I passed the MEPS physical.

198006: USAF Law Enforcement poster in the recruiter's office, June 1980

I arrived home around lunch and waited 24 hours to tell my parents. They were shocked, to say the least. My mother started shrieking about how I would get killed, and suddenly relented to letting me enroll at any college I desired if I would drop the idea of enlisting. I replied that it was too late, that she should have supported my earlier desire to enroll at UT-Martin. Besides, I had already signed the contract — another revelation that caused great gnashing of teeth and more wailing. What annoyed me most was her insinuation that I was rebelling against her authority and letting down our family by not acquiescing to her vision of my future. I would never escape her bipolar mood swings and matriarchal domination unless I left and chose my own path in life.

That night, my father came upstairs to talk again, asking me if this was what I absolutely wanted to do. I said that it was. He replied, "Then I guess we have no choice but to accept it." As he left, he turned and added, "For what it's worth, I believe you will do quite well." Drafted in 1950 and having served three years in the US Army in Germany, he knew I could handle it. Word got out quickly that I had enlisted. Few were surprised and almost everyone supported my decision. Most warmly congratulated me, especially all the older relatives. My great-uncles, who were WWII veterans, talked to my mother and helped calm her down about it. Their collective opinion was that it wasn't wartime and I'd probably get out in four years to attend college using my veterans' benefits. They reasoned that by that time, I would be mature enough to balance sports and academics.

Yet some thought otherwise. One former teacher predicted to a number of people that I'd never make sergeant, much less attend college. This news got back to my parents, and it especially upset my mother, a fellow teacher. My father said, "One of my teachers said something similar. I obviously proved her wrong. When I graduated from college, I mailed a graduation invitation to her. Remember this so you can do the same someday." Over the years, the former teacher lived long enough to see me become a sergeant, earn four college degrees culminating in a doctorate, and retire as a lieutenant colonel with almost 31 years of service. But in 1980, all this lay far ahead.

I got on the Greyhound bus to Nashville for my MEPS physical late Wednesday afternoon. I was greeted at the Nashville bus terminal by a big Marine NCO (sergeant) who directed me to another bus that would take me and a load of others to a nearby hotel. Another NCO assigned me to a room with another guy who was enlisting in the Army. That evening a lot of guys got really drunk. I had a couple of beers, but otherwise, I just sat around talking to other guys who were destined for the various services. I bought a bottle of Calvados (apple brandy from Normandy) at the liquor store next door to take back to the farm.

The next morning, they woke us up at 0430 and herded us down to breakfast. One clown who had gotten drunk hours earlier vomited in the garbage can by the front door as we loaded into the bus. The Army sergeant escorting our group bawled him out and put him on the bus anyway.

Upon arriving at the MEPS building, we had to fill out a pile of forms before getting checked out. The really drunk guy was in the next row and two seats behind me; he barfed all over his form, and they sent him away. I was glad to see that foolish guy go.

Our next stop was a big room where we stripped to our underwear. One by one, different doctors came by to check us for every possible thing. Several very amusing events occurred, making everyone except the examining doctors laugh. During the rectal exam, one guy moaned out, "Oh, doc. I thought you didn't love me anymore." We all cracked up, and the doctor ignored it while the NCOs yelled at us to shut up. They also checked our testicles and this spawned more sexual-themed remarks. The funniest was when a doctor asked someone why he only had one testicle; the guy replied, "Sir, I lost it from a football injury," which drew even more laughter from the crowd. During the screening questions about homosexuality, one guy asked, "Is being a fag required? Because I really look forward

to serving at sea!" The NCOs constantly snarled warnings that were mostly ignored since this craziness went on all morning. I was glad when the MEPS physical finally came to an end. By 10:30, I was back on the bus and en route home. That evening, I sat on the front porch with our Collies for a long time, puffing on my grandfather's pipe to drive away the mosquitoes. So many things were running through my brain, and I was thrilled to have set my future in motion.

I settled back into farm work, fishing, playing in the summer softball league, and coaching my brother's softball team until my basic training date in September. I also served as a counselor at the 4-H camp in Columbia, Tennessee, for a couple of weeks and attended the 4-H Roundup at UT-Martin. I bid farewell to my friends over the summer and prepared to take on bigger things. I checked back with the recruiter in Florence to pick up my in-processing packet and plane ticket the week before I left. Everything was set.

On 24 September, my father dropped me off at the Nashville Airport. He was unusually quiet on the trip up and hugged me as we said goodbye. An hour later, I boarded a plane with several other guys heading to Lackland AFB. No one talked much.

We had a two-hour layover at Dallas-Fort Worth Airport. One of my travel companions, Joe Steele, was from Huntsville, Alabama and had identical orders to mine, so it appeared we were in the same basic training unit. We decided to hit the bar and have a few daiquiris while we waited. That would prove unwise, as we soon discovered. We boarded our final flight to San Antonio feeling pretty good.

We were greeted in San Antonio by several Air Force NCOs who directed us to a large holding room. One of the NCOs welcomed us and advised us to hang tight. A short time later, we were directed to board a waiting bus. Joe and I sat together in the back row. As we loaded up, everyone was excited, jabbering on about the usual things:

where they were from, women, sports, and what specialties awaited them. The driver was a sloppy looking, overweight E3 who stepped on the bus and yelled, "Sit your sorry asses down and shut the hell up! Welcome to the United States Air Force." Joe and I started laughing. The driver turned around and snapped, "You won't be laughing an hour from now, you assholes." We yelled at him to go fornicate himself, and he shook his head. We kept laughing, figuring we wouldn't laugh much in the weeks ahead.

A little while later, the bus turned off the interstate and passed under a big overhead sign that read: "Welcome to Lackland Air Force Base – Gateway to the US Air Force." We had arrived at our new temporary home. The bus pulled up in front of a low building, and we were ordered to get off. Drill sergeants (Training Instructors or TIs) in dark blue Smokey Bear hats started barking orders at us from the start. "Get in line! ... What the hell are you looking at? ... Quit eye-balling me, boy! I'm not your damned girlfriend! ... Eyes front, shit bird! ... Stand still! ... Stop fidgeting!"

I knew that the "artificial stress" of basic training had officially begun! This ended quickly as we were rushed inside to a large room, where we filled out forms before being separated into our basic training squadron groups. The TIs pointed out a big amnesty box into which we were instructed to discard all illegal items: weapons, pornography, tobacco, alcohol, and other contraband. I snickered at one guy who dumped in a stack of Playboy magazines and another who shoved in several small bottles of booze. I wondered why they bothered bringing all that stuff to basic. Joe and I were assigned to Squadron 3702, Flight 398. The reception NCOs briefed us on "dos and don'ts" before calling us each by name to get on buses taking us to our squadrons. And off we went again.

By now it was dark outside and after not eating since the layover in Dallas, we were getting hungry. After our warm greeting at the reception station, we remained silent on this short trip across the base. The bus stopped in front of an enormous building with cavernous overhangs that resembled what my mind imagined was a parked spaceship. A young NCO ordered us to fall out, grab our bags, and get into formation under the overhang. We stood in silence for a few minutes before hearing the clacking sounds we would never forget — the approach of our TIs.

Two TIs walked briskly toward us and began bellowing, "Pick 'em up! Put 'em down! Pick 'em up! Put 'em down!" The old "pick 'em up / put 'em down" game had commenced. Upon order, we picked up our suitcases and held them for a moment, followed by an order to put them down. The orders grew faster and guys started dropping luggage; this incurred a personal dogging by one or more TIs. Those who didn't move fast enough were dropped for pushups and inevitably we all got dropped.

More TIs showed up and the mind games intensified. One heavy-set guy didn't move fast enough and a TI yelled, "Get down and beat your face!" The kid dropped and began hitting his face against the concrete, resulting in the TI bellowing, "Are you damned stupid? I'm talking about pushups, you dumb bastard! Start pushing Mother Earth before I rip your head off and shit down your throat!" Yep, this was definitely what I expected. Within a minute, we all were doing pushups to pay for the slow ones. And we did hundreds.

After a while, the TIs grew bored with this routine and all but two disappeared. The senior one introduced the TI team. "My name is Staff Sergeant Spence, and that's Sergeant Brentham. You will speak only when addressed and follow our orders to the letter. The first and last word out of your ugly holes when you speak to us is 'Sir!' Do you understand?"

In unison, we screamed, "Sir, yes, sir!" The TIs were apparently deaf as hell because we had to repeat this several times and did dozens more pushups. Tiring of this domination game, they ordered us to ground our gear (stack it in a neat pile) and file into the chow hall to eat. The games continued as they taught us how to receive food and eat. Everything was a facing movement and so on. We had about two minutes to choke down all we could and get back under the overhang.

The TIs took us upstairs to our flight barracks. It was divided into two large open bays with a large training room and latrine room at the west end. On the north end, below the ceiling was a line of windows too high for us to look through. The TI office, with one bunk, was located next to the stairwell door. Joe Steele and I bunked next to one another; on the other side of me was a Georgian named Delaughter. We dropped our bags and were herded into the training room for our welcome briefings. We filled out more forms and a postcard to send home to let our loved ones know our new mailing address.

The TIs introduced themselves to us in turn. A college graduate, SSgt Spence was leaving for Officer Training School in a couple of weeks, so we were his last basic flight. I don't recall anything else about him. Sgt Brentham had not been in the service long (five or six years, I think) and was up for promotion soon. He made it clear that he was our master, not our friend. His constant swearing in creative, unusual combinations was shocking to those who had never been exposed to harsh language before.

After collecting the postcards, Sgt Brentham looked around the room. "Which one of you is the biggest asshole?" No one moved or said anything. Looking at my Nike track shoes, he pointed and said, "You look like an athlete. You're now my latrine queen. Your job is to make sure the latrine is spotless before we start training each day. If it's screwed up, it's on you. Pick five guys to assist you, any five you want." Although I just met them, Steele and Delaughter volunteered along with three others. I knew immediately I didn't like Sgt Brentham and wouldn't get along with him.

After another hour or two of briefings, they let us make our bunks and stow our gear in our wall lockers. Thirty minutes later, Sgt Brentham cut the lights. He had a bunk and a small latrine in a small room at the front of the bays. It was 0200.

At 0500, the lights came on, and SGT Brentham's screaming voice had us leaping from our bunks. "Get the hell out of those racks, shitheads!" A big steel garbage can crashed along the floor between the bunks. Everyone was standing up at attention except Delaughter, who was still snoring despite the noise. Brentham grabbed his bunk and hoisted it up against the wall with an instantly awake Delaughter now trying to scramble out of it. I lost it and started laughing. Brentham yelled, "Twenty pushups! Get down and beat your face, asshole!" He went on down the line, harassing and insulting the rest in the two squads sharing our bay.

We stood there in our underwear for a while as the TIs found some reason to mess with everyone. Finally, they had us get dressed to fall out for breakfast. We hustled downstairs and went through the chow line taking anything the cooks put on our trays. Bacon, powdered eggs, grits, and toast were standard fare. The Yankees didn't touch their grits and the TIs tore into them about wasting food. Seeing the TIs make them eat grits was hilarious but I didn't dare laugh. With breakfast over, we were back upstairs for a 100% shakedown inspection. We laid out everything we had brought with us on our bunks. A female first lieutenant walked by my bunk and casually looked it over before moving on. One guy brought a box of condoms with him and the TIs descended on him like angry wolves. One TI said, "Boy, the only ass those are getting put in is yours!" Among the TIs was MSgt Oakley, the squadron training NCOIC. He was a loud Arkansan who missed no chance to harass us, and he carried a big knife that he liked to whip out and brandish to frighten the weak trainees.

After the shakedown inspection, we marched to the barber shop and base supply. The barbers cut off everyone's hair to make everyone look alike. One guy had two braided ponytails that reached almost to his waist; the TIs kept his hair as a souvenir. Another had an enormous Afro. The barber carefully removed it in a single piece and then stood it on his shelf like a helmet. One white kid showed up with a shaved head; a TI whom I had never seen before placed the Afro on the bald kid's head for the barbers' amusement. The distraught looks on so many faces about losing hair was amusing since I didn't see it as a big deal. We walked out of the barber shop looking like a bunch of bald scrotum heads and that's exactly how Sgt Brentham described us.

Supply was next and getting our uniforms seemed like it took all day. As we were sized for uniforms and boots, fat civilian clerks threw items at us to cram into our duffel bags. Finally, we got everything and marched back to our barracks. The TIs showed us how they wanted everything displayed in our wall lockers and later took us out for some physical training (PT).

As expected, they discovered I was a runner — fastest in our group — and appointed me to the squadron track team. From that point on, the TIs were a lot nicer to me than to my buddies.

We did a lot of marching and close-order drill during the first week. We must have yelled, "Sir, yes sir!" a thousand times, along with our march chant, "3-7-0-2, Flight 3-9-8!" We had a few slugs, but they were balanced by sharp guys, some with junior college degrees. One of our sharpest individuals already had a bachelor's degree and two years of experience in Air Force ROTC. He enlisted to "gain some experience and perspective" before applying for Officer Training School.

On my seventh day in basic, SSgt Spence called me into his office. He said my records reflected I had three years of JROTC and was eligible to bypass the last half of basic training. This would allow me to report to SP School early and possibly enter an earlier course. He explained it was not automatic, and I had to pass a series of written and practical exams. Two others in my flight were also approved to pursue the bypass option — Bird and another's name, I cannot recall. The college graduate and one other JROTC alumnus decided not to apply. The next day, we bypass options were called to meet MSgt Oakley. He made it clear we had to max everything to bypass. So from that point on, whenever the flight went out to march, we were diverted to his office to study, take exams, or do work details for the squadron.

One negative was that bypass options were often stuck in a waiting area with the screw-ups and quitters bound for the SLL ("sick, lame, and lazy") barracks. Some SLLs would bide their time complaining or crying, so we wanted to kick their tails to shut them up. None seemed to have any viable reason for wanting out, just couldn't hack basic training. We were glad to see them go since we didn't want to serve with them. Anyone unable to handle basic training was unlikely to handle real life, much less combat operations. In reality, basic and advanced training courses are designed to create artificial stress, both to build individual resiliency and to teach coping skills. Weeding out the softies and slackers early saves taxpayers' money and ensures the military has a reliable, mission-capable force. One day, MSgt Oakley overheard me tell an SLL to shut up with the crying before I slapped the hell out of him. He invited me into his office for what I expected to be a tail-chewing; instead, he gave me an overview of how our basic training squadron was organized.

Latrine duty was easy. To make cleaning easier, we blocked off half the latrine stalls and banned soap in the showers. Instead, we bought several large containers of shampoo and insisted that everyone use it as a body wash to reduce soap scum. The other trainees were happy to comply since it meant their soap dishes stayed clean and dry in their locker drawers. The TIs acted surprised that we could get the latrines cleaned so quickly, yet left us alone about how it was done. I never once had to pull kitchen police (KP) during basic training.

On our second day, SSgt Spence asked if any of us could type, and I was the lone person who could. Instead of getting KP as an additional duty, I served as a charge of quarters (CQ) clerk in the squadron admin section. Every third evening, I worked a 4-hour shift, mostly typing reports and answering the phone. It was an easy duty that I didn't mind at all, compared to sweating over dishes and mopping floors in the chow hall, or raking grass clippings and picking up cigarette butts outdoors. As such, I saw MSgt Oakley and the squadron command team members frequently.

We had a guy with a difficult-to-pronounce Serbian name, so everyone called him Consonants. He seemed like a great guy and was glad to serve. One morning, he suddenly decided he'd had enough of basic and declared he was a homosexual. The TIs didn't believe him and hauled him down to see MSgt Oakley. I was outside the latter's office and could barely contain my laughter as I listened to the interrogation unfolding through the large window and adjacent door. It was cordial at first, but then it grew loud.

MSG Oakley yelled, "Airman Consonants, are you honestly going to stand there, look me in the eye, and lie to me about being queer?"

The young man replied, "Sir, I am a flaming homosexual, completely gay, and hereby request an immediate discharge, sir."

"That's bullshit! I don't believe you!"

"Sir, I'm telling you, I'm a homo and queer as hell. That makes me ineligible to serve.

You have to discharge me and send me home."

"Son, I don't believe you. How long have you known you were gay?"

Consonants was silent for a moment and said, "When I got here to basic, sir. I can't control myself in the shower. I get excited when I see the other guys naked."

MSgt Oakley started to laugh and then grew stern again. He stood in front of him and said, "I still don't believe you. Get on your knees." He then unzipped his fly as if to pull out his willie and said, "I guess you'll just have to show me to prove it."

The kid had a horrified look on his face but he recovered well. "Master Sergeant Oakley, I'm honored by your very sincere offer but you're just not my type."

MSgt Oakley bellowed, "Get the hell out of here and back to your flight. We'll decide what to do with you."

Consonants grinned as he walked by me. I suppressed my return smile as soon as I saw MSgt Oakley walking through the door. He motioned for Sgt Brentham to follow. "Start processing him out," he ordered. Glancing in my direction, he said, "I'm not saddling a future commander with a loser like that." I nodded my head in agreement and went back to typing. By the time I arrived back upstairs to our barracks, Consonants was gone — already sentenced to administrative purgatory at the SLL barracks.

On that topic, there were several barracks past the main base exchange (BX) that housed airmen who were convalescing or awaiting discharge for various reasons. Base leaders kept them isolated far away from other trainees. I think the convalescing airmen were further segregated at that site based on whether they might reenter the training pipeline once healthy.

At the end of my second week, we bid farewell to SSgt Spence, and Sgt Brentham was our lone TI. The following week, we three bypass options spent almost no time at all with the flight. We had our squadron track meet that week, and I easily won the mile run with a time of 5:33, greatly pleasing our commander, first sergeant, Sgt Brentham, and MSgt Oakley. My reward was getting to eat several ice cream sandwiches during CQ duty that evening.

At the start of the third week, Sgt Brentham told us to bypass options to report to the obstacle course the following morning at 0600 and to the rifle range that afternoon. Passing these was a mandatory basic training requirement. I knew we would easily pass both. The obstacle course was a couple of miles away, and we marched over to it at dawn.

Because bypasses were not part of the other training flights, their TIs made us go through first. The Public Affairs Office (PAO) team had a group of reporters there conducting a story, so we were required to provide our names and hometowns. They took a lot of photos and videos of us going through the course; when the photographers didn't get the right shots, we had to repeat a few obstacles. The morning dew was thick and covered every obstacle, so our uniforms soaked up it and the dirt from crawling under barbed wire. Before we reached the final obstacle, we looked like muddy hogs. The PAO team loved how we looked but I sense the photographers were disappointed that none of us fell into the water traps interspersed throughout the course. Not to worry, they got their wish with the flights behind us.

We marched back to the barracks to change uniforms and eat lunch before going to the rifle range. Since it was over by the adjacent Kelly AFB, we were taken there in a truck and dropped off; we'd ride back on the bus with other flights firing that day. We picked up M16s and ammo, received the training and safety briefings, and were put in the first firing order. I hit 100% of my targets and qualified as an Expert Marksman, as did my two bypass option buddies. When we returned to the barracks, MSgt Oakley was very pleased and thanked us. Since he knew we were all Southern farm boys, he bet another NCO ten bucks we would all score Expert.

Midway through the third week we bypass options had passed all our tests and were told we would report to our next assignments (tech schools) the following Monday. We were allowed to have our Airman First Class (A1C) stripes sewn on one uniform each, but we could not wear them until the morning we left for our tech schools.

That weekend Sgt Brentham drilled us hard and harassed me in particular every chance he had. He got in my face several times, yelling about what a screw-up I was and how he would not allow me to bypass basic training, no matter what my scores said. He got so close he spat all over my face as he screamed at me. He finally grabbed me by the throat and screamed, "Hit me! Go ahead, I know you want to! Go for it!" I figured it was some sort of test so I took the abuse from this smaller man whose only advantage was his rank and authority. But deep down, I wanted to punch him out over that tirade.

On Monday morning, Byrd, the other bypass, and I showered and put on our uniforms with stripes on the sleeves. We said goodbye to our buddies during breakfast, then packed our gear and went downstairs to out-process. Sgt Brentham came by and said, "Y'all are still screwups, but best of luck in tech school and beyond. Don't get anyone killed." And with that, our nemesis walked out of our lives.

I toted my suitcase and duffel bag over to the Security Police School five blocks away and reported in at the 3781st Training Squadron. I was assigned to a "casual" barracks with others awaiting their start dates at tech school. By bypassing the second half of basic, I reported to SP School three weeks ahead of schedule. A fact not previously shared with me was that I would wait for my originally scheduled SP class to begin, rather than joining the next available one. This meant working details around the base — tasks like KP, CQ, garbage hauling, grass mowing, and other menial maintenance chores. Fortunately, the reception NCO was happy to discover I could type and assigned me to CQ duty at nearby Camp Bullis. I knew this was a Texas Army National Guard installation on the northwest loop around San Antonio and hosted the USAF SP Air Base Ground Defense (ABGD) school.

Casual duty was a breeze. I had each day off until 1800. After chow at 1700, I caught a ride with the mail shuttle to Camp Bullis. My duty was to man the CQ office, answer the phone (which never rang), and make hourly security checks around the tent city occupied by tech school and ABGD students. Security and ABGD students were trained at Bullis for two and nine weeks, respectively, whereas Law Enforcement students (like me) spent only a week there. I was allowed to take catnaps during the midnight shifts and had weekends off, so CQ was a good gig. The Training NCOs I worked with at Camp Bullis were friendly and shared a lot of useful knowledge with me. They certified me to drive the M151 jeep and several other vehicles, and annotated my training records; the benefit was that I reported to my first assignment fully qualified to operate multiple vehicles. I also toured all the training areas and ranges, and thus received good info on what to expect in tech school. Another cool thing about CQ duty there, instead of at the main base, was that the NCOs let me eat ice cream crunch bars from their refrigerator.

After duty ended each morning, I caught the outgoing mail shuttle back to Lackland AFB and sacked out until late afternoon. Unfortunately, I was awakened constantly by others in the casual barracks. After two weeks, I was told to move to another barracks and join my training flight. So I packed up my gear and humped it four blocks away to my new home.

I joined Delta Flight — the "Delta Dogs." Our Training chief was MSgt Tholen, and he supervised three other NCOs. The only one whose name I remember is SSgt Rodriquez; she was a great instructor but had a nasty temperament, very volatile even toward honest mistakes. MSgt Tholen was a big fellow and nice to us for the most part. I was assigned to a two-man room with Steve Borden, a former dock worker from Tampa, Florida.

Our days began at 0415. We woke up, did the "three Ss" (shit, shower, and shave), cleaned the barracks, and marched off to breakfast. By 0630, we were in class and trained until 1730 when we broke for evening chow. Sometimes we trained again after chow. This went on for the next eight weeks. We had most weekends off and spent some of our free time in a nearby snack bar and dance hall. I went running every chance possible to stay in shape. Our PT sessions were generally easy and mostly consisted of pushups, sit-ups, grass drills, and short sprints. Whenever we received the order to march, we alternately stomped our right feet as we yelled on our left step, "D Flight (stomp) Total Force (stomp) Blood!" Every training flight had a unique slogan or "sound off" call. While they meant nothing to anyone else, they let anyone within earshot know who was in the training area.

Early in our SP training, Ronald Reagan was elected President. Most were happy about it, particularly our instructors. Congress had already passed an 11% across-the-board pay raise for 1981, a vote many felt was designed to help re-elect President Carter. President Elect Reagan announced he would seek more military funding to deter the Soviet Union's aggressive expansion and break it financially if possible. While no one considered that we might see an end to the Cold War anytime soon, Reagan's election seemed a step in the right direction.

One Monday morning, we had an inspection, and we stood at parade rest in the hallway while it took place. Using strings, I had rigged a uniform in my big wall locker so that when the door opened the gloved arm rose up in a salute. I heard my inspector mumble, "Holy mother of God. You gotta be kidding me." He then summoned all the other instructors to come try out this feat of textile engineering. I could hear the instructors inside snickering as they opened and closed the wall locker over and over. Suddenly, I heard the window open and knew the unmistakable sound of my footlocker, bedding, and uniforms getting thrown out the second-floor window onto the ground below. I did at least 100 pushups for that stunt. Yet thankfully, our room passed inspection.

I enjoyed getting mail from home and wrote to family and friends whenever possible. All wanted to know where I would go after SP School. Of course, everyone was anxious about our future assignments, and most got their assignments around the fourth or fifth week. The Sunday after Week 5, I was eating Chinese food, and my fortune cookie read: "Soon you will cross over the Great Water." My friends all said I was heading overseas. I had signed up only for US bases along the Redneck Riviera so figured I was heading to a base in Florida, Georgia, or Alabama. The next day, I received my assignment: 3rd Security Police Group, Clark Air Base, Republic of the Philippines. Instead of the Southeastern US, I was going to Southeast Asia! My parents were not thrilled, and I sensed my mother sorely regretted not letting me enroll at UT-Martin.

Ten guys total from Delta received orders to the Philippines, including my roommate, Steve Borden. The others were Bruce Bielby (Bethany, Missouri), Todd Crocker (Fayetteville, North Carolina), Mark Denesia (upstate New York), Wayne Griffiths (Indiana), Steve Boudah (Utah), and a dual-citizen Briton named Samuel Smith. Two married guys also went to Clark AB: Ray Ashbrook, a married Chicagoan, and Kelly Owens, whose wife (also named Kelly) had enlisted and would join us in the Philippines after she completed training. Denesia, Crocker, and Smith met and married Filipinas so they and our other married buddies would remain in country for three-year accompanied tours.

On weekends, we put on our Class B uniforms and rode the bus to town. Mostly, we hung out at the River Walk in downtown San Antonio. As long as you ordered food or drinks, you could sit in the cafes and watch the tourists go by in the small river shuttles. We all visited the Alamo and explored other local sites of interest. The Alamo's curator was aware that the only life-size statue of Colonel David Crockett was in my hometown square in Lawrenceburg. While home on Christmas leave after SP school, I mailed him photos of it.

On some Sundays, I attended chapel services with other airmen from the tech school. Incidentally, many basic trainees were taken to the chapel by their TIs; the latter usually went elsewhere during the services and returned afterwards to retrieve their charges. Although there was a chapel next to the SP Academy, the chaplain there was a "fire and brimstone" speaker. So, mostly, we went to the one farther away, where the female basic trainees went.

One Sunday after chapel services, a 6-foot female TI (whom I'd have found quite attractive in another setting) came to pick up her flight. We stood outside under the awning, and she eyed us suspiciously. Having had a chaplain talk openly to them in a no-threat environment, the female trainees suddenly emerged from the chapel laughing and cutting up. The TI ordered them to fall in and launched into a profane tirade. Pointing at us, she declared, "Y'all see those men? There's over a mile of dick on this base and you won't see one inch of it until AFTER you graduate. Now when I say 'forward march' I don't want to hear anything but 40 pussies clapping together in unison and your boots striking the pavement until we get back to the barracks! Forward, march!" I started laughing, and the TI turned to glare at me. "You shut your mouth, boy!" Correctly figuring her attention was focused on her trainees, she kept marching on. I noticed the chaplain (a major) standing in the chapel doorway, shaking his head at her in disgust.

One evening while enjoying music at the snack bar dance hall, I noticed several tall blondes walk in. The tallest one made eye contact with me, and I asked her to dance; Led Zeppelin's Stairway to Heaven started, and she was fine with a long, slow dance. Christine Debevec was an interesting Minnesotan whose older brother was a chief master sergeant. She had a two-hour pass from her basic training unit and was scheduled to graduate the following week. Although four years my senior, we had a great time talking, and I sensed our attraction was mutual. We swapped addresses and were pen pals until she was assigned to the Philippines a few months before my tour ended. I was supposed to meet her on the River Walk during her town pass that weekend; however, I was with another group of friends, and we never met up.

Another night at the snack bar, I lost track of time and realized I'd miss curfew if I didn't run back to my barracks five blocks away. I sprinted across an open drill pad and jumped over the ditch next to my barracks. Another man was sitting outside on the front steps, smoking a cigar; he yelled something at me that I couldn't make out, and then screamed, "Skunk!" Running too fast to stop or alter course, I leaped over a skunk that had sauntered around the corner of the barracks next to the smoker. The surprised skunk sprayed the area as both of us scrambled inside and slammed the door. We looked at each other, shrugged, and headed to our rooms. Within minutes, I could hear guys yelling out and slamming their windows down. Our barracks smelled like a skunk for several days.

During Week 6, we were awarded our dark blue SP berets, the tech school rite of passage. I shaved mine to remove the excess fuzz, wet it down, and shaped it, then put it on. I continued to shape it until it was dry, and then pinned on my Pacific Air Forces crest. Naturally, my classmates took photos wearing theirs to send home, and I was no exception.

198011: Law Enforcement class photo, Nov 1980. I was in third row at center above class leader wearing red shoulder rope.

Weeks 7 and 8 consisted of law enforcement-specific training, as well as night maneuvers. The final week and a half, we trained at Camp Bullis and its weapons ranges. It was December, and it was cold sleeping in the Group Medium tents. Temperatures dropped into the 20s every night, and we had no cold-weather gear other than field jackets (without liners) and gloves.

Something occurred at Camp Bullis that caused me to generally avoid resorting to group punishment for the rest of my career. One morning, after we were transported to our training site, the instructors discovered that three trainees had left their gloves in their tents back at the camp. So to enforce uniformity, the rest of us were forbidden to wear our gloves despite the freezing cold and sleet that day. More than 20 trainees suffered from mild frostbite just to make a foolish point about uniformity. To my great disappointment, the instructors wore their gloves, thus setting a piss-poor leadership example. I promised myself that — barring a valid justification — I would never use group punishment in such scenarios, especially those that might cause undue risk of injuries to others.

With the wind howling out of the north every night, our small coal heater did little to warm our tent at night. We were allocated only a small bucket of coal, so we used it sparingly. However, the heaters were enough to warm up our C-rations; these were often unappetizing and even more so when almost frozen. I warmed up my C-ration cans for the next day's meals until too hot to handle without gloves, then shoved them into my sleeping bag to warm it and keep my feet warm. I quickly discovered no one else liked the "Ham and Eggs, Chopped" meal, so I traded for that meal all week. Many other C-ration entrees were heinous at any temperature. One of our Mexican-American guys had a big bottle of Texas Pete hot sauce, and that sauce made the entrees taste much better. Thereafter, for the rest of my three-decade career, I never went to the field without a bottle of hot sauce.

We had our only Sunday afternoon and evening off, so we enjoyed a few cold Coors, also known as Yellowjackets. It sleeted or snowed several days in a row that week, though there was no real accumulation. During our capstone exercise, the instructors took special care to throw artillery simulators into our bunkers to force us to exit into the open; we made such easy "kills" for the opposing forces. We still enjoyed the training, especially the time spent at the ranges for weapons fire. When our training at Camp Bullis ended, we trucked back to Lackland for a final day to process out before graduation.

I recall very little about graduation day on December 23, other than being surprised by how many of the guys' families traveled there for it. A general was our guest speaker, and I remember nothing he said except "best of luck to you and serve well as we counter the communist threat." In my mind, I heard Obi-Wan Kenobi saying, "May the Force be with you, always." Late that afternoon, I rode a bus to the San Antonio Airport and recall sitting in the front row, watching snow flurries hit the windshield as the traffic zoomed by. A few hours later, my mother picked me up at the Nashville Airport. It was good to get home to the farm and sleep in my own bed. I had 30 days of leave at home and would then head out on my next adventure in a foreign land.

Christmas with my family was both enjoyable and depressing. Our family was never the same after my maternal grandfather died two years earlier. I walked into the large implement barn where he suffered a heart attack in 1976; it was mostly empty of equipment, and I hated its very existence. Moreover, I still felt sick over losing his presence in my life at age 16, as he didn't get to see me or my siblings play high school sports, graduate, or grow up. Had the barn not stored the last few remnants of our now-idle farm equipment, I might have burned it down. To this day, I still have negative sentiments toward that barn.

Knowing I was going away for the next 18 months, I visited as many relatives and friends as possible while home. My brother Shaen was playing middle school basketball for EO Coffman Junior High, and I attended several of his games. I went deer hunting a couple of times and intentionally didn't bag any so I could enjoy the quiet time alone in the woods. My sister visited a couple of times from college, but we didn't hang out since she had her own circle of friends, including several whom I intensely disliked.

Among the first things I noticed was how young most of my high school friends seemed. To be sociable, they asked about my training and where I was going next, usually the extent of the conversations. Everything I was about now was alien to them, so they quickly lost interest and always changed the subject. It seemed most were completely unaware of the Soviet threat or that the Cold War even existed.

My parents were vocally worried about me going to the Philippines. They heard all sorts of stories about wild living, communist insurgents, and other potential dangers. I kept telling them I'd be fine and not to worry. I was grown now and trained to do my job. No matter what I said, they were going to worry until I returned home again.

The next weekend, we visited my father's family near Huntsville, Alabama. All were very supportive of me enlisting, especially my paternal grandmother. My great-grandmother was 92 and still mentally sharp. She was proud of me and asked that I honor our family's heritage. Her grandfather was a Confederate officer in the 19th Alabama Infantry Regiment; after receiving a crippling wound while commanding his company at the Battle of Chickamauga, he made it home to Alabama and resumed his Methodist ministry. Her father-in-law had served as a cavalryman under Captain Truss in Hilliard's Legion, an independent Confederate cavalry unit. Her sons had served in World War II, and then several of her grandsons and nephews served in Korea or Vietnam. I was the only great-grandson in uniform, and this greatly pleased everyone; to my knowledge, I remain the last of her descendants to have served in the US military. My uncle Tillman, a Korean War veteran who worked at Redstone

Arsenal, picked at me nonstop and predicted I would get thrown in the stockade at my next assignment.

After the holidays, my final three weeks home were uneventful. My college friends headed back to their campuses, and everyone else returned to work. Life was moving on for everyone, and I grew stir-crazy. Repairing fences and other farm work helped the time pass. Our Collies followed me everywhere. Their lives seemed comparatively simple: eat, sleep, bark, poop, bark at the cows, chase cars, bark some more, and follow family members around the farm. I honestly envied them despite their relatively short lives.

Finally, my last day at home arrived. A few relatives and friends stopped by to wish me well. I packed up my gear, loaded the car, and posed for a few photos taken by my mom. My father and Uncle Rex (his brother-in-law) drove me to the Nashville Airport and waited until my plane to San Francisco left. Back in that era, there were minimal security checks, and guests could wait with flyers at the gates. It was 22 degrees with snow flurries, so that made the sunset glow on the western horizon. I was soon swallowed up in the belly of the big aluminum-skinned bird and off to my next adventure.

"Zoomie Pig Odyssey, 1980-84"

Philippines, 1981-82

Assigned overseas for my first duty station in the US Air Force (USAF), I received orders to report to the 3rd Security Police Group (SPG) in late January 1981. We were given a load of advance reading material from the Security Police Academy library, and I researched much more about my future home. I also received a sponsor letter from the unit, along with a welcome packet containing several informative brochures. One thick brochure was from the 3rd SPG, and another was from the base public affairs office. Although I was obviously apprehensive about traveling so far, I was also very excited to go. Ten guys from my class were going there, and a couple were military brats who had lived there before. Of our 10, two were married, and the rest of us were single.

The 3rd SPG provided security and support to the USAF within the Republic of the Philippines (aka the Philippine Islands, PI, or RP), an island archipelago in the western Pacific Ocean, located just north of the Equator. The nation was plagued by low employment, communist insurgents, and corruption within its own government. It was a largely underdeveloped country with a diverse terrain—mountains, dense jungles, rolling agricultural lands, and stunning scenery. It had a long monsoon season (May to October) and frequent typhoons, offset by a warm but mostly comfortable dry season. Aside from the sprawling capital metropolis of Manila, 70 miles to the south, there were few large cities, and most "urban" areas were just small towns, much like those found throughout the rural United States. As agriculture was the largest domestic activity, most Filipinos ("Pinoy," as they preferred to be called) were businessmen, farmers, fishermen, craftsmen, and transporters, among others, who lived in small towns and villages near their farms and trades. Overall, the Pinoy were friendly, loved talking to us, and welcomed our presence.

Clark Air Base (CAB) was located in Pampanga Province, in the central part of Luzon, the largest northern island. It was initially founded as Fort Stotsenberg in 1901 to house cavalry units within the newly established American colony. Soon after, the US Army built Clark Field a couple of miles southeast to base the first biplanes in the country. Over time, the Army post and base merged into a single larger installation. Japan invaded the Philippines in late 1941 and occupied it until the US liberated it in 1945. After granting the country independence in 1946, the US maintained control of Clark Field, then greatly expanded it over several decades. Following the catastrophic damages to base facilities by the Mount Pinatubo eruption in 1991, US forces ceased operations at CAB and turned it and the nearby Subic Bay naval complex over to the Philippine government in 1992.

The Philippine Air Force (PAF) had around — we guessed — 75 total fighter planes, with 20 or so based or rotated through CAB and the rest distributed to other bases. While the Philippine Army (PA) had several units stationed near us in various locations to assist with security, most were based elsewhere. The PA had a few dozen UH-1 "Huey" and other helicopters with most scattered throughout the archipelago. Interestingly, the Singapore Air Force had a half dozen fighter planes stationed at CAB; its support personnel lived in a barracks near the Base Consolidated Armory (BCA). Most Pinoy officers spoke decent English, as did many enlisted men. Most schools taught English and television stations had limited American programs. While most newspapers were in Tagalog (the main Pinoy language), some were published in Tagalog and English.

The historic 13th Air Force had operational command and control of all USAF units in the southwestern Pacific region. Its headquarters graced the western end of the old Fort Stotsenberg polo field and parade ground; the 3rd Combat Support Group (CSG) headquarters occupied the eastern end, and the CAB Officers Open Mess ("The CABOOM") was on the north side about midway. The parade ground featured large trees, wide sidewalks, and a number of historical monuments. Monthly retirement ceremonies took place at the west end, with the base honor guard presenting formal salutes. Surrounding the parade ground were several housing areas, mainly the old "barn"-style homes elevated on stilts to keep out snakes and rats. Chambers Hall (bachelor officer quarters) occupied the northeast corner.

The main US operations element was the 3rd Tactical Fighter Wing (TFW), which hosted two squadrons of F-4 Phantom fighters and KC-135 refuelers, plus the 21st Tactical Fighter Tactical Aggressor Squadron (TFTAS), 3rd Aerospace Rescue and Recovery Squadron (ARRS) — aka Pararescue — with its Jolly Greens and rescue helicopters, and multiple other units. The base had sufficient operations and parking ramps to handle the 3rd Tactical Airlift Wing's (TAW) large air cargo fleet: three C-5 Galaxy, around 15 C-141 Starlifter, and 20 or so C-130 Hercules turbofan planes. There was also an Aeromedical Evacuation unit that had several C-9 Nightingale transports. At the west corner of the C-5 ramp were the alert hangars, typically occupied by F-15 Eagle fighters rotating on temporary duty (TDY) from Kadena AB, Okinawa. Aside from a few other hangars and "barns" to shade the mechanics, most aircraft were parked along wide rows on asphalt parking ramps.

Tucked within support buildings around the taxiways were various asphalt work and repair pads. A few were on semi-grassy spots covered with perforated steel planking (PSP) or "Marston Mat" left over from World War II. Immediately to the north, a long row of 20 barracks built around 1960 was laid out in six diamond shapes of four each. These ran parallel to the ramp areas, so each large unit's barracks were generally perpendicular to its facilities, shops, and hangars. Six additional modern barracks were built along the next boulevard to the north, housing personnel from other support units. Two additional barracks complexes (built in the 1950s) occupied either side of Lily Hill; most on the north side were dilapidated and abandoned. In the mid-1980s, the latter were refurbished and reoccupied.

The 3rd CSG was the primary support element for the 3rd TFW. Under it was a large assortment of support units, including food services and housing, engineers, a hospital, personnel, finance, fuelers, legal services, chapels, intelligence, training, and security. The greatly appreciated Morale, Welfare, and Recreation (MWR) team included all sports facilities and leagues, as well as clubs (Airman, NCO, and Officer), and local sightseeing tours.

Further under the 3rd CSG administrative umbrella were all the maintenance, supply, and communications units. These included the 3rd Field Maintenance Squadron, 3rd Equipment Maintenance Squadron, 3rd Aircraft Generation Squadron, 3rd Transportation Squadron, 3rd Supply Squadron, and other units. The 1961st Communications Group handled worldwide communications, to include electronics security and satellite relays; it also managed the giant "elephant's cage" regional surveillance array on the north side of the base.

Armed Forces Radio and Television Service (AFRTS) ran the Far East Network (FEN) radio station. We always laughed at the acronym AFRTS because even its employees pronounced it as "A-farts." However, FEN mostly played music, so we enjoyed its presence. Every hour in the afternoons had one hour dedicated to various music genres, including disco.

There was also a US Embassy contingent at Clark AB. Their dependents lived in base housing and attended the DoD Dependents Schools on base. We rarely saw the embassy teams, except when they entered or left the Regional Relay satellite communications center, located between the Main Gate and Dau Gate.

While the 374th Military Airlift Wing belonged to the larger Military Airlift Command (MAC), it was attached to the 3rd TFW and provided airlift support to the entire southwest Pacific area of operations. These MAC flights went worldwide, and there was a steady stream of flights in and out of Clark AB. It was possible to catch a "space available" flight anywhere in the region—Japan, South Korea, Hawaii—or back to the US. Virtually all "Space A" travelers took leave because there was always a risk of getting bumped by higher-priority passengers (those traveling on TDY orders) and stranded at a faraway destination until the next military flight. Sometimes people took Space A flights to their destination and flew commercial back to the Philippines.

My advance literature went into great detail about my new unit. Under the 3rd SPG were the 3rd Law Enforcement Squadron (LES), 3rd Security Police Squadron (SPS), and the 3rd SPG Administration (SPA). Under the 3rd LES, there were four law enforcement (LE) flights/shifts designated A–D and the Traffic/Motorcycle patrols; these had security responsibility for the non-flying portions of the base, also known as the "cantonment" area. The 3rd SPS had four Security flights/shifts, K9, Horse Patrol, and additional resource protection assets; these assets had security responsibility for the flight line, runways, and areas beyond, including perimeter security. The 3rd SPG SPA fell under the Command Group team and oversaw Pass and ID, Records, Crime Prevention (SPAC), the SPG Motor Pool, Town Patrol (a.k.a. Quad-Agency Patrol), Security Police Investigations (SPI), Detention Center (base jail), SPG Training, Animal Control, and Correctional Custody. The SPA also included the Base Merchandise Control section that sought to prevent black marketing. Notably, horses could access terrain that vehicles could not, providing an invaluable capability; the only other horse patrol section in the USAF was located in Panama. Likewise, the 3rd SPS Military Working Dog (MWD or K9) section had around 200 dogs and was (I believe) the largest K9 section in the world. The K9 guys liked to refer to their section as the "Clark County Sheriff's Department."

The base was roughly a 5-square-mile bloated triangle (sort of liver-shaped), with a 10,000-foot runway and a parallel taxiway that ran nearly the same length. The perimeter fence was roughly 28 miles long, with perimeter observation towers every 250–400 meters apart, depending on the terrain. There were a few concrete machine gun bunkers or "pillboxes," including one or two screening the end of each runway. Back from the runways, there were a few mortar pits to allow indirect fire support in case of attack by insurgents. The latter positions were left over from the Vietnam Era and were in disrepair. While doctrine called for wide clear zones stretching 25–100 meters past the perimeter fences, in some places, local farmers grew crops right up to the fences or walls. The only walls at the time were on the northwest and part of the west sides, stretching from base housing down past Sapangbato Gate. In many other places, the barbed wire fences were gone—stolen by thieves to sell on the black market. Base engineers (mostly contracted Filipino civilians) frequently bush-hogged or hand-cut the grass during the monsoon, both inside and outside the perimeter, to keep vegetation low for easy intruder detection. During the dry season, either the SPS, K9, or PAF guys just burned the expanses of grass once, and the ground stayed bare until the next monsoon arrived six months later. Of note, this was a wonderful experience for the pyromaniacs.

The 3rd SPG patrolled the perimeters and manned a few towers at night, while its K9 and Horse Patrol units were constantly on the move to deter or capture intruders. While the PAF was supposed to provide manpower support to the security operation, such support was rare since it struggled just to maintain its aircraft and flight operations. That's how my unit and similar units stayed busy at CAB. While the US officially remained neutral in the conflict between the Philippines and its communist rebels, the US concurrently would not allow another allied government in the region to collapse and go communist. Meanwhile, despite hosting US forces, the Philippine government's formal policy denied US forces to engage hostile locals except in self-defense. Obviously, the PAF and the Pinoy Army were more than happy for the help, even if it meant US forces occasionally going on offense, which both our and their commanders would categorically deny, and we never discussed while in-country.

In late January, my month-long leave at our Tennessee farm ended, and I was off to my next adventure. My father and Uncle Rex took me to the Nashville Airport for a night flight to San Francisco. It was cold—22 degrees with snow blowing sideways—yet I was sweating underneath my Class A uniform. I don't know whether it was from nerves or carrying two bags with everything I was allowed to take overseas. I could tell my father was nervous about me heading to a foreign country, particularly one with a culture completely alien to anything he had experienced during his US Army years in Europe. Of course, people had warned about all sorts of nefarious activities rumored as the norm for servicemen in the Far East. I'm certain these things crossed his mind as we sat in the airport restaurant drinking coffee. Uncle Rex was a Navy pharmacist's mate in Guam in the late

1950s, so I suspect they discussed plenty en route home after dropping me off. Eventually, my boarding call sounded, and I waved goodbye as I checked into my gate.

Hours later, I stepped off the plane in San Francisco. I ran into a couple of other guys in uniform at the baggage carousel; most were heading to Japan, Korea, or the Philippines. It was midnight, and a tired-looking USO rep herded us together and motioned us to a waiting bus bound for Travis AFB near Oakland. A drunken Navy NCO in the back served as an unsolicited tour guide as we passed by landmarks. We were grateful when he finally passed out, but then annoyed because he snored like a gargling elephant.

The reception center at Travis AFB assigned me a room for the evening. The other bunk was occupied by a large Black fellow who stirred when I walked in. I undressed in the dark as a courtesy and hit the rack. The next morning, I awoke around 8:00 to the sound of the shower and met my temporary roommate, Danny Elum, a transportation guy also heading to the Philippines. He was a college graduate who enlisted when he failed to land the job he wanted. It was 61 and sunny, so we ditched our jackets and reveled in the warmth. We headed to the bowling alley for lunch and met up with several others bound for Clark Air Base. Steve Borden (my SP Academy roommate) and Bruce Bielby (another classmate) were among them, so we joked and poked fun at one another to pass the time. The day passed quickly, and that evening we boarded a 747 Flying Tigers jet with a couple of hundred others destined for Clark or points along the way.

The flight to the Philippines was far longer than I had anticipated, and this was my first time ever seeing the Pacific Ocean. Winging into the evening sky toward our first stop in Alaska, I had a window seat in the nose and could see the Cascade Mountains pass by to the east. After a couple of hours, darkness overtook us, and I could see nothing but an occasional light below. I spent most of the time reading a Newsweek magazine I picked up at the Oakland terminal. One of the guys next to me traded seats with someone behind us, and Steve joined me across the aisle. Before long, Bruce Bielby traded seats and joined us, as did another from our class (Greg Truskowski), who was getting off the plane in Okinawa. A while later, my ears began popping as we descended into Anchorage Airport.

As it had for Vietnam veterans and thousands of others before us, a huge stuffed polar bear greeted us as we debarked at the Anchorage Airport, where we met up with several other classmates. Now a posse, we grabbed sandwiches at the restaurant and magazines to read on the longer flight legs ahead. After taking off again, I finally dozed off and slept until awakened by dim light through the window. It seemed like forever before the flight attendants brought us food, which consisted of a baked chicken rear quarter, biscuit, apple, and a Coca-

Cola — standard airline fare at the time. I often looked out the window and was captivated by the endless rippling blue ocean below. Finally, an island came into view, followed by another, and then white-topped mountains pierced the wispy clouds through lush green forests. I correctly guessed we were passing over northern Japan. Before long, clouds enveloped the mountains, and that was the extent of my view for several hours until just before we landed at Kadena AB, Okinawa. It was pouring down rain, so we saw nothing but planes on the tarmac as our plane disgorged part of its human cargo and gulped in a few more. In 30 minutes, we were airborne again on the final leg of our flight.

The ocean south of Japan was bluer than anything I had seen before. I'd remembered reading that the Pacific Ocean was its deepest there. I spotted a few tiny specks far below, obviously ships. Before long, I made out green forested mountains ahead: the island of Luzon. Once over land, the terrain was steep, rugged mountains and patchworks of small fields fanning out from valleys into the flatlands. As we descended, I noticed the dry rice paddies were dotted with black spots; I later learned farmers burned the rice stubble for the fertilizer value, to kill insects, and decrease rodents. Rice and sugar cane fields dominated the wide Luzon plain that was intersected only by roads, riverbeds, and towns along them. Cars and buses moved along them at a snail's pace, and my mind wandered mutely about the scene unfolding before me. We were now over what I had read was

Pampanga Province. We were low enough that I could easily make out ant-like individuals below, and judging from the acreage of the cane field, sugar cane was the dominant crop.

Glued to the window, I saw a large airfield ahead at the base of a range I knew was the Zambales Mountains. As we grew closer, the base perimeters were readily apparent from abrupt lines where buildings stopped and open space began. Vast swaths of blackened terrain stretched from the perimeters to the runways and taxiways. I was impressed by the size of the base as well as the expanse of parking areas filled with rows of large cargo planes and dozens of F-4 Phantom fighters. Behind these were large hangars watched over by a sentinel control tower and hundreds of buildings stretching beyond into the foothills of the mountains. All this quickly raced by, and I glimpsed the South China Sea beyond the mountains for a few minutes as the plane flew past the airfield in a large arc to approach from the southwest. The airplane turned sharply, and I saw a mostly dry riverbed that I soon learned was the Abacan River. I now noticed carabao (water buffalo) and farmers in the fields, and the perimeter fence zoomed past as we passed overhead. As we touched down, several people behind us cheered — apparently guys who were returning home or had been here before and were glad to be back.

198110: Flightline photo of F-4 Phantoms being "generated" for missions during Operational Readiness Exercise in October 1981. The powerful twin-engine F-4 fighter was proof that even a brick could fly if given enough horsepower thrust.

Next to the nose door, I was shocked by the heat as it opened to my new home. Although January is winter in the Philippines, I was not acclimated to the 75-degree temperature and began sweating again. Permanent party personnel and dependents returning from leave or vacations stateside were ushered straight into the MAC Terminal. Our reception team herded the rest of us into an adjacent building with a tin roof; I thanked God that I was next to an open window. We filled out a pile of forms and received half a dozen briefings — a few I would remember — before a Public Affairs major walked in to formally welcome us to Clark AB. Following this, a large sergeant named Jones announced that he was there to pick up all the new 3rd Law Enforcement personnel. We loaded our gear into a blue van (converted bread truck) for a short ride to our barracks complex to sign in at the 3rd LES orderly room.

Our barracks sergeant — aka barracks manager — was an unfriendly type who made this clear from the start. I later discovered he was friendly; our treatment was standard fare to keep us on our toes the first few days. We received more briefings and were told we were under post-restriction for the first 72 hours "for your own protection." I assumed this was the case until later that day, when several NCOs came by to say they were taking us on a local tour at 1800; one of them said to eat at 1700, so we had full stomachs. When I protested that the barracks sergeant had put us on restriction, one laughed and said, "Well, I outrank him and y'all are coming with us. Be ready to go in civilian clothes when we arrive at 1800 sharp."

I drew linens and a key to my barracks room. I inadvertently woke up my roommate, an older buck sergeant named Dennis Keefe, and could tell it upset him since he worked midnight shifts. I changed clothes as quietly as I could, ditched my gear in a wall locker, and headed out to roam around. A couple of us looked up SGT Jones, who graciously took us for a ride around the base in the van so we could take notes on where things were for reference. While a large installation, most places we needed to go were within walking or biking distance. For the rest, taxis were available for reasonable fees. I made a note to myself to buy a bicycle.

Our barracks were somewhat substandard by continental US (CONUS) standards, yet decent enough to include hot water and air conditioning that worked most of the time (emphasis on "most"); some guys also used fans to pull air through the screened windows. Barracks Row and the runway areas were just high enough in elevation (480 feet above sea level) that nights cooled off rapidly, especially in the dry season. The high Zambales Mountains to our west shaded the base from the late evening sun, allowing us to enjoy glowing sunsets over the peaks. Most barracks were surrounded by large trees and had wide porches shaded by overhangs from the rooftops. Housing areas and barracks alike typically had large trees, beautiful tropical shrubbery, and immaculate lawns.

Every barracks had local "houseboys" assigned to do laundry, shine boots, sweep floors, and maintain the barracks and lawns. It was a mutually beneficial arrangement since it provided lucrative employment to them and freed us to work, rest, or recreate. Their services were Monday–Friday and cost $20 per month. This was worth every penny, especially if your barracks had a large lawn like ours did. While an occasional lost item was expected, the houseboys were generally honest and rarely took anything. I always tipped my houseboy, named Rey, 50 pesos per month (approximately $ 5 USD) or gave him gifts that he could take home. As subservient as it might sound, a houseboy taking care of 20–25 GIs earned a fairly lucrative salary (before tips and gifts) in a depressed Philippine economy that had 30% unemployment at the time.

Although the barracks had two-man rooms with screened windows, most common areas were open air. We only closed the central doors and aluminum window louvers during typhoons. As such, flies and mosquitoes flew through the open common area doors and always managed to get into our rooms. As a general rule, we didn't bother the geckos and other lizards that also got in because they ate the flying insects. Flies ruled the day, and mosquitoes ruled the night.

I found the chow hall and ate a big meal — free with my meal card. Most guys living in the barracks ate in the same mess facility, and the food was good. Breakfast was standard fare (sausage/bacon, oatmeal, or cereal with eggs, toast, and coffee or fruit juice), whereas the other meals consisted of pork, chicken, or fish with rice and a local fruit (mangoes, papayas, or guavas) or some dessert made from sugar beans and coconuts. We also had some beef shipped in from Australia, although not every day. Whatever rice didn't get eaten on a particular day, we received the next day either mixed with fruit or included in the dessert dish. My favorite was the rice pudding, despite knowing the rice was made with leftovers. Occasionally, the 3rd CSG would fly in steaks — a special treat.

Milk was the shelf variety or reconstituted with water from a dry powder, which ranged in taste from OK to heinous, depending on its refrigeration. One could pour in a scoop of sorghum molasses and pretend it was a milkshake if desired; in a country full of sugar cane, sorghum was dirt cheap and used to flavor many desserts. While the Yankees turned their noses up at sorghum, many Southerners grew up eating it, so they were fine with it.

Ultimately, those who worked hard and played sports got hungry and would eat anything. Of note, if you had a good friend in the 374th TAW, it was possible to talk one into bringing back fresh milk from flights to Japan. The one time that happened for me, it tasted funny, so I gave it to our houseboys, and they loved it.

Of note, our milk came from the Philippine Area Exchange (PHAX) Dairy. It was a group of buildings located by the Base Mortuary out near the riding stables. They mixed dry milk powder with water and maybe some canned condensed milk to create a fair — yet by no means good-tasting — milk substitute. The running joke was "the PHAX Dairy is the only dairy on the planet with no cows."

Unit commanders did their best to encourage sports and fitness, yet it was nearly impossible for most units to schedule fitness training due to rotating shift work. The various intramural sports events were intense and highly competitive. Commanders loved having winning sports teams for bragging rights (friendly wagers, too) and frequently had their strongest athletes assigned to day shift jobs to guarantee availability for night and weekend games. However, like anywhere else on the planet where free time, money, and boredom collide, off-duty GIs would gang up and head "downtown" for fun. The extensive bar district supplied plenty of entertainment, and local police had their hands full keeping order at times. Most bars were rather tame and offered standard entertainment — jukeboxes, drinking, pool, darts, and girls dancing on mirror-backed bars. Some bars and clubs were rather raunchy, and a few downright hedonistic, catering to bizarre tastes best left to one's imagination. And whatever you can imagine … go lower … then go much lower.

Desperately poor people will do about anything for money, and Filipinos were no exception. The positive aspect is that roughly half the single men assigned to CAB found wives this way, so it was a win-win for all involved. Whatever experience one desired, it was available along the main drags of Fields Avenue and MacArthur Highway or at any number of venues on the side streets. General MacArthur would have rolled over in his grave had he known his namesake highway was widely referred to as "Blow Row." Ultimately, every large city in the US has similar bars and clubs, but in developing countries, it is easily affordable to average young American guys.

True to their word, at 1800, a couple of NCOs picked us up at the barracks, and we headed downtown. En route to the main gate, we received the "unofficial" briefings about things that could get us into big trouble. We were warned to tread lightly and get the feel of things before saying or doing much. As we pulled up to exit the main gate, the LE guard looked at us suspiciously; he could tell we were new guys, and he stared at us with a stern demeanor. As he handed our ID cards back, his face broke into a big grin. He said, "You boys have a great time tonight. Welcome to the PI, mother f—kers!" We all laughed as our car drove away.

Immediately outside the gate was a wild, heathen world full of bars (some little more than strip clubs and cat houses) and every vice one could imagine. We pulled into the parking lot of a "cop bar" and went inside for a few beers. I had heard about places like this

— bikini-clad women dancing on the stage, dirt-cheap booze that flowed freely, and older guys with exotic girls hanging all over them. Obviously put up to it by our NCO hosts, girls sidled up to us to declare how we were the most handsome guys they had ever met and to ask us to buy them drinks. We escaped the latter because none of us had any pesos yet, so the NCOs bought them for us. I thought I was here to help protect your country, so why don't y'all buy ME drinks? I also learned a new term that evening: "Cherry boy." Regardless of one's previous sexual experience, marital status, or even orientation, every new male arriving in-country was a cherry boy. I heard guys argue vigorously that they were not, only for the girls to insist they were until they had "mated" with a local girl. So, unless you had a Pinay wife or girlfriend, you were automatically considered a cherry boy by some, including our PAF gate guard counterparts. The crazy part was still getting called that through the end of my tour, I suppose, because I looked much younger than my actual age.

One observation from my arrival at Clark and first foray downtown was the unique smell. The area smelled unlike any place I had ever been. Depending on the time of year, it was a mix of smoke, various foods cooking, raw sewage, garbage, vehicle exhaust fumes, mown grass, livestock shit, and whatever plants were blooming. And with a climate similar to South Florida, something was always blooming. During the dry season and post-harvest crop field burnings, the air stayed smoky for weeks at a time. While the odor was less obvious on base, it grew more intense the closer you got to Angeles City and other cities in general.

Another thing that struck me as odd was how so many drivers honked their horns while driving. They seemed to do this at random for no reason other than to make noise. Over time, I asked drivers why they did this, and none could satisfactorily explain it other than claiming everyone else did it. It made for a noisy ride or walk anytime you ventured downtown.

Like other nocturnal creatures, food vendors increased exponentially at dusk to sell their wares to those partaking in the local nightlife. While a few vendors had jeepneys outfitted as food trucks complete with grills, most had simple charcoal grills set up on the ground next to the streets. The NCOs took us to a vendor they knew, and we sampled the pork sticks on bamboo skewers. They warned us to get to know vendors before buying food, since food poisoning was a common threat. Some vendors bought discount meat that was already going bad or had never been refrigerated. Plus, GIs also risked getting dog, cat, monkey, or some other mystery meat. Usually, I politely declined when friends bought from vendors to avoid risking food poisoning.

The NCOs told us there were over 200 bars in Angeles City, mostly along Fields Avenue and MacArthur Highway. At any given time, there were at least 5,000 bar girls registered with the local health department. Together with a base hospital team, the public health department routinely checked each bar's girls once weekly for sexually transmitted diseases and common illnesses. Anyone diagnosed as unwell was prohibited from working or even serving drinks. There was an inspection schedule, and each bar had a registry card for each girl working there; health officials and police could check these at any time and shut down a bar for violations. While some commanders cringed at this practice, it was a logical tool to prevent STDs that could cause potential readiness losses. It was a lot to mentally process on our first night.

Breaking the restriction was our informal orientation to how most single guys spent their free time in this wondrous new place. Funny thing, the girls warned us to go slow on the San Miguel beer, as there was sometimes poor quality control on alcohol content. Where you might drink several and feel little effect, other times one beer was strong enough to knock out a carabao. We hit several bars along Fields Avenue and one on MacArthur, so several hours passed by in a flash. After a while, the NCOs hauled us back to the barracks and dropped us off. As the others chattered on about all the women and partying ahead, I felt bewildered and alone. While unfortunate in some ways, this initial orientation jaded my outlook on this assignment for weeks until I ventured beyond the immediate area to see more of the country. Welcome to the P.I.

This was the strangest place I had ever been in my life. The culture was indeed more alien to anything I expected, less so as time passed and I grew comfortable with going places. I figured out that the microcosm around the base represented both the absolute best and absolute worst of the country. However, for the most part, I found the Filipino people warm, friendly, and extremely hospitable. Most over age 40 remembered the horrors of the Japanese occupation during World War II; many openly voiced and displayed gratitude for their liberation decades before. Some displayed both Philippine and US flags at their homes and businesses, especially the WWII veterans.

The older Pinoy loved to talk to Americans and share stories. They wanted to know everything about you — where you were from, what you did before enlisting, whether any of your relatives had visited or served in their country before. These people would give you the shirts right off their backs, offer to share their food, and then apologize for not being able to give you more. It was a very poor country by US standards, and most Filipinos toiled greatly to make a decent living. I made it a point to take food and other small gifts when I went off base, especially when visiting a Pinoy or a local family. With only a few apple orchards in the country, apples (whole or dried) were a very appreciated gift.

Conversely, criminal elements were drawn to the military bases, attracted by the prospect of a more comfortable life than working in agriculture or services. The crooks often targeted American servicemen whose comparative wealth, bad habits, and naivety frequently combined to make them easy prey. If you crossed the wrong individual, you could get beaten up, stabbed, or even killed — same as most any other country — so it was wise to travel with a wingman or in groups. At times, combating criminality and vice seemed forlorn at best, a race between education and catastrophe.

Among the first things we were warned about was black marketing. Many BX and commissary items were controlled and rationed. We received ration punch cards for the latter, which the BX would scrutinize at checkout. Consumable items, such as food, liquor, and cigarettes, were controlled on a time basis; one had a reasonable amount of time to consume them, and in the meantime, was expected to account for them. You had to register motorcycles, bicycles, and all electronic gear with the Merchandise Control Office (MCO), managed by the 3rd SPG. The MCO tracked both incremental and large purchases to combat black marketing; if identified as a potential violator, the MCO would investigate and figure it out. If caught black marketing, one would lose BX privileges, sometimes permanently, depending on the severity of the offense.

Most Americans were honest about their buying and stayed within their limits. If hosting a party or event that requires extra food or spirits, it's easy to get permission to exceed your limit. The key was to request it in writing well in advance. Many like me who didn't smoke still bought their monthly cigarette ration and then doled them out judiciously to friends or houseboys as gifts. This was common, and the MCO wasn't concerned about it. However, when someone was identified as potentially selling exchange items for profit, the MCO often paid a visit for a "show and tell" inspection. I knew several non-smokers who got caught selling their cigarette rations on the black market and paid a stiff price to make a few bucks.

On that topic, the black market was a thriving business off-base. Downtown, beyond the bar district, was an economic oddity known as the Nepo Market. Ironically, this was

"open" spelled backward. You could buy all sorts of things sold in the BX (food, household items, clothing, disposable diapers, etc.) at marked-up prices. You could also buy replacement ammo and even weapons for the right price. Early in my tour, I bought a couple of M16 rounds of every lot number in our armory after dropping an M16 magazine; a single round popped out and bounced down a storm drain. I not only paid a dollar for the lost round, but spent a day off getting signatures from various base agencies to affirm that I hadn't sold the damned thing. Following the wise suggestion of a senior NCO, I prevented any future occurrence by buying a replacement stash of various lot numbers and burying it next to the barracks. Before I rotated stateside, I traded my bullet stash to a friend for a cheeseburger and fries at the Airman's Club.

For the most part, in-processing into the 3rd SPG was uneventful. Among our first briefings was a very detailed task organization overview. Every Air Force unit in the Pacific south of Alaska belonged to Pacific Air Forces (PACAF) and was mutually supported by airlift units from Military Airlift Command (MAC) and fighters from Tactical Air Forces (TAC). The 13th Air Force commanded all units and elements in the Southwest Pacific area of operations (AO); its HQ was on the west end of the base parade ground. In the Philippines and surrounding airspaces, the 3rd TFW controlled all air operations, whereas the 3rd CSG managed all non-flying support operations. The 3rd SPG fell under the 3rd CSG and supported both it and the 3rd TFW. The 3rd SPG was divided into three parts: 3rd Law Enforcement Squadron (LES), 3rd Security Police Squadron (SPS), and the SPG HQ element or SP Administration (SPA). The 3rd LES oversaw law enforcement operations — police flights (gate and patrol shifts) and traffic control (motorcycle police). The 3rd SPS oversaw flight line security, K-9 operations, and horse patrol. The HQ element included SPA, Customs, Training, Supply, Pass and ID, Merchandise Control, Quad Agency Patrol (Town Patrol), SP Investigations (SPI), Crime Prevention, and Resource Protection. To simplify command and control, the 3rd SPG had just reassigned most Resource Protection posts to the 3rd SPS and K-9, so it was down to a skeleton crew doing special details.

As we processed through finance and other agencies, we attended various briefings in large groups. We met the 3rd TFW and 3rd CSG commanders, respectively, Colonel Corder and Colonel Max Weiner. Colonel Corder's portion was quick: "Ours is a big mission, and I'm glad you're here, so do your duty to keep our planes ready for combat." Colonel Weiner's welcome was more genial and relaxed; he offered us valuable tips for success and promised that we would see him often. He was true to his word and frequently appeared at base and unit functions. He tended to ride his moped around the base instead of driving his staff car, often with a big unlit cigar clenched in his teeth. I definitely liked Colonel Weiner's style.

During in-processing, we were also issued our DEROS dates — Date of Expected Return from Overseas Service. Mine was in late July 1982 unless I was granted an extension in theater or got married and had dependents present. The latter created an accompanied tour status and an automatic extension to 36 months. Many singles requested extension anyway, and some did their best to serve their entire 4-year enlistments at Clark. Upon receiving orders, barracks occupants would tape their orders to their doors with the acronym FIGMO (Finally I Got My Orders) inked across them. The acronym for those getting out was GTFH (Going The F—k Home). One generally received follow-on assignment orders six months before the DEROS date.

The most interesting aspect of in-processing was meeting the 3rd SPG commander, Colonel Gary G. Allison. A former Army enlisted man with tours in Vietnam for both the Army and Air Force, he was a towering presence. A Texan with a gold-plated cow turd on his office wall, he was direct and spoke his mind with clarity and force. Prior to his command tenure, the 3rd SPG was short of nearly everything and had older vehicles and equipment, much of it in dire need of replacement. When he arrived in mid-1980 (then still a lieutenant colonel), he submitted requisitions for shortage items and additional items he contended were mission essential. He reportedly changed the rules of engagement (ROE) to one of "If fired upon, return fire and then radio in that you are receiving fire." He was adamant his SPs had a right to self-defense and would not die waiting for someone on the other end of a radio to grant permission to fire back. Colonel Allison immediately set higher performance standards, and morale improved exponentially. His direct leadership style and acidic humor were intoxicating to SPs accustomed to boring commanders who would drone on about global mission priorities, values, platitudes, and esoteric stuff; Allison explained things in simple terms that everyone could understand. He hated illegal drug use and conducted a series of no-notice drug screenings, then kicked those who tested positive out of the 3rd SPG. In fact, that's why so many LE airmen — 10 from four straight SP Academy classes — were arriving in country at the time.

198101: Colonel Gary G. Allison, 3rd Security Police Group Commander, Clark Air Base, Jan 1981

This commander directed us to speak clearly using plain English, insisting that any reasonable person should be able to regurgitate back exactly what we said. He said, "If I visit you on a gate and the phone doesn't work, I don't want to hear 'Sir, I perceive some difficulty with the ringing mechanism' or other bullshit like that. Unless you're a phone technician, you won't know. Just say, 'The phone is f—ked up.' This tells me you know exactly what's wrong with the item, so I can submit a work order for someone to fix or replace it. This is what I want from you — accurate and timely information that lets me provide you the right tools so you can accomplish your assigned duties and our unit's mission." He went deeper in detail as to why no information was better than reporting inaccurate information.

Allison also introduced several leadership concepts that served me well throughout my career. "Your leaders are more responsible TO you than FOR you." He explained how each one of us filled a specific niche within the organization and that his headquarters existed to support us in both peacetime and wartime roles. "Y'all are my most critical asset. Whereas I can replace machines and equipment by reallocating resources until new ones arrive, replacing people and accumulated knowledge is significantly more challenging. If your leaders are properly training and supporting you, we'll work as a team to destroy any enemy stupid enough to engage us. But if they aren't, we risk defeat at multiple levels." He continued, "Engagements and battles are won or lost at the fire team and squad levels — not by armies and certainly not by grandiose, esoteric shit barfed out by faraway generals and politicians. Just like in football, I need you to beat the asshole in front of you, then the one behind him, and so on. Once we kill enough of the bastards, they'll quit and go home or leave us alone to go bother someone else. War is a brutal and ugly business, so if you doubt you have the stomach for killing other men as a political act, tell me now so I can get you reclassified into another career field. I won't tolerate cowards, dopers, and lazy SPs."

Then came the summary. "Do your best at every task every single day. I'm not asking you to be Superman, just to do your level best and earn your pay. As General Lee told his men, 'Do your duty always, for you cannot do more and should never aspire to do less.' And one more thing: Don't ever lie to me about anything. It's not like when you screw up there's some big guillotine in the sky that swings down to chop your heads and peckers off. Just tell me the truth, and we'll figure out how to fix it. Aight?" We were pretty excited about serving under this commander, and after he dismissed us, that's all we talked about over lunch. This guy was already a legend in SP circles, and we felt more confident knowing he was in charge.

Colonel Allison often convened monthly Commander's Calls at the NCO Club, where he paid for kegs of beer from his own pocket. He took questions directly from the audience so his troops could hear the answers firsthand; he had someone tape these events so he could follow up on unresolved issues. One time, when someone asked the same recurring question about supply problems, Allison's answer got a standing ovation. "Look y'all. I don't have a bunch of garden gnomes sitting in a back room eating raw materials and money and shitting out uniforms, boots, and equipment. I've ordered all this shit and will have the Supply and Training teams issue it out as soon as it arrives. So in the meantime, I need y'all to be patient." As the laughter died down, the poor airman still stood there, as if frozen. Our commander graciously told him to sit down, explaining to the audience that it was an honest question and deserved an honest answer. But the laughter was thunderous.

Commanders greatly desired championship sports teams, so Colonel Allison had the Training Section identify those with athletic skills as they arrived. He had these SPs reassigned to day shifts or administrative jobs to make them available for his sports teams during evenings and on weekends. During exercises, he eliminated the requirement for guards having to stand on empty tarmacs to guard "simulated" aircraft; he told the wing commander that unless an aircraft was present to guard, he would "simulate" the SP guarding it. He drove around the flightline constantly during exercises to check on the welfare of his SPs, especially during aircraft generations. He ensured that patrols closely monitored guards for signs of dehydration and heat injury, and he physically checked each post to confirm that everyone had eaten during meal times. During one exercise, I was on a far perimeter post when the colonel himself brought out boxed meals and cold water to the guards. He explained that this gave him the opportunity to monitor his people and assess the effectiveness of his security operations. As a result, we enlisted men worshipped our leader, whose radio call sign was "Lone Star." Whenever there was a problem, it was reassuring to hear a voice drawl, "Lone Star to Command Post," over the radio.

Although I didn't know it when I arrived in-country, I had two NCOs watching over me from a distance. One was SSgt Osteen (3rd SPG Horse Patrol), whose cousin, Ronnie, worked at the Lincoln County (TN) Farmers Co-op and was friends with my father. The other was SSgt Stewart Venable (Base Engineers), whose sister, Beverly, was in my high school class. Another from my class, Airman Jeff Polk (374th TAW), arrived several months after me and was then sent back to CONUS for an undisclosed issue. I never did find out what it was.

I was grateful for the "life guidance" from the Pinay wives of my colleagues and friends. While some initially hoped I might find a local girl and settle down, once I explained my career and education goals, the vast majority urged me to remain single. Several then openly "forbade" me from getting into a serious relationship, mainly because I was young and had plenty of time for marriage and family after college. One flatly told me, "You'll marry much better if you're an officer." She added, "You marry the entire family, not just your wife." I did notice that many American officers with Filipina or other Asian wives tended to marry college-educated women. While none of this really entered the equation at the time, it reinforced my desire to remain single until after I completed college.

Concurrent with the usual in-processing tasks, every FNG ("f—king new guy") went through at least two weeks of local indoctrination training. During it, we were issued jungle fatigues and other theater-specific gear. MSgt Don Funk was the Training NCOIC and had around 30 FNGs in training, including the 10 from my SP Academy class. I learned a great deal from the Training team NCOs, most of whom were either Vietnam veterans or on their second or third tour with the 3rd SPG. Funk brought in experts from all over the base to brief us on various theater-specific topics — weather, heat injuries, snakes and venomous insects, venereal diseases. His reasoning was simple: Threats often turned into readiness losses, and prevention is almost always cheaper than a cure.

198102: Staff Sergeant Donald Funk in the bomb dump at Clark Air Base, January or February 1981.

Among the serious threats I learned about and later experienced were thieves, insurgents, and snipers. Because the latter could not get close during the day, their occasional potshots were usually wild and missed. The PAF perimeter tower guards would sometimes return fire at night, but not always, due to the risk of hitting a friendly or getting shot themselves. That also assumed the tower was manned at all, since, by my estimate, 95% of the time, it was vacant. Once night fell, the thieves and intruders (along with their lookouts) would come in closer and infiltrate the base at dozens of points. Mounted horse patrols and K9 teams on patrol were a fairly effective deterrent, especially during the dry season after all the elephant grass and vegetation had been burned off. Horse patrol and K9 were considered dangerous and stressful duties, requiring strong dedication and steady nerves.

Our SPG elements sometimes teamed up with the Department of Defense (DOD) contract guards who patrolled at night. If fired upon, the ROE was to return fire immediately, but let the senior leader on duty (typically a shift commander, usually a captain) decide whether to develop the situation. In reality, everyone thinking straight would fire back at the muzzle flashes. These little engagements were infrequent but occurred often enough that one stayed alert on patrol. That said, there was a much higher probability that a thief would just sneak up and swipe your weapon, radio, or gear if left unattended or in an unlocked vehicle. As in any sudden close encounter, the ROE could easily get ignored, as startled or scared young guards were somewhat unpredictable. More often, there were shooting incidents involving DOD contract guards. These were mostly Ayta tribesmen, whom the Spanish colonials called "Negritos" or little brown ones. The Negritos were indeed short, with dark skin and curly hair — mountain people with a long heritage of opposing foreign forces or anyone who showed them disdain. They had an array of primitive weapons and skillfully used them as they lived "with" the land instead of just on it. The Spanish and Japanese discovered that the Aytas were experts at insurgency and feared them. Already familiar with them from his previous Philippine tours, General MacArthur employed their services as scouts and raiding parties after invading Luzon in early 1945. For their valued contributions, the US Government granted the Aytas medical privileges for 30 years at CAB. Working on base after that afforded (albeit limited) continued access to medical care, so many did their best to secure on-base employment.

A large number of Aytas and mixed ethnicity Filipinos lived just off base in Negrito Village (later renamed Marcos Village). King Alfonso was their chief and a World War II veteran, so I was advised by older NCOs to salute or bow to him when he came to the gate. The man was maybe 5 feet tall, spoke fluent English, and seemed very friendly. After the first time I saluted King Alfonso, he took a liking to me and always asked about my family in Tennessee. While he also appreciated that I brought the village kids candy, he asked me to curtail this since the sugar was bad for their teeth and diets. He was fine with me bringing fruit from the chow hall, though. Alfonso looked after his young people in that way, so I really liked this fellow.

The 3rd SPG had a couple of hundred DOD employees, including many Ayta guards who carried Remington 870 shotguns (12 gauge) that were nearly as long as they were tall. That said, in a criminal or intrusion confrontation, the Ayta guards were more likely to gut or decapitate the offender using their favorite sidearm — razor-sharp machetes. I had to guard a crime scene one morning after a private Ayta guard killed an intruder in an officer's housing area. Allegedly, the guard leapt on the thief from a tree and took off his head when the intruder fought back. The intruder's body was headless, and it was never clear where the head went. I didn't ask and really didn't want to know.

I had heard plenty of stories about the Negritos prior to my arrival, so wisely made friends with them. Once you earned their trust, they watched your back. I also let them "salvage" whatever they wanted from the dumpsters and only lightly inspected their loads of items as they exited the gate. They especially liked aluminum cans, since they could reuse them for many things. Some rooftops in Negrito Village were very clearly "shingled" with flattened soft drink and beer cans.

Funny thing, before I left the country, I sold my wrecked 10-speed bike as salvage to an Ayta fellow for 5 pesos (at the time about 50 cents) and had to produce a bill of sale to comply with Merchandise Control rules. I wasn't allowed to "gift" it to him.

We had plenty of slap flares for emergency nighttime illumination. These were small white phosphorus flares inside an aluminum tube that fired like a mini mortar when struck against a solid object to ignite the rocket propellant. The flares rose a couple hundred feet and made a distinct pop as they ignited, then burned for 15 or so seconds as they floated down on two-foot-square parachutes. Flares produced a low hiss and a weird, glaring light that made objects appear flat yet moving at the same time.

Night patrols typically carried three to six in a shoulder bag or secured them to their load-bearing equipment (LBE), and these were fine to light up an area. The problem with flares was that they ruined night vision; thus, it was wise to use them with your shooting eye closed to preserve it. The other issue was that they drifted on the wind and could ignite dry brush or rooftops, so we had to be careful using them. Conversely, if you wanted to burn an area of brush to deny intruders a place to hide, they were ideal—just fire the flare sideways and enjoy the show. Guys with pyromaniac tendencies discovered this right away.

And I certainly torched my share of grass and brush once the dry season began.

Aside from the 3rd SPG's bragging rights as the largest SP unit in the Air Force, a surge of new SPs was assigned there due to a major drug bust a few months earlier in 1980. Illegal drugs and thievery were Colonel Allison's pet peeves, so he and his executive officer (XO), Lieutenant Colonel Peter Quist, cleaned house. They busted a wide range of folks—patrol cops, K9, and customs—for a lot of different offenses. Naturally, many squealed and ratted out their buddies for their participation or other crimes, so this turned into an expansive purge of miscreants, malingerers, and malcontents from within the 3rd SPG. Among the worst was a Customs NCO caught trying to sneak a football full of heroin into Japan; he served part of his prison sentence in Japan and the rest at Fort Leavenworth, Kansas.

When our commanders lacked sufficient evidence to prosecute offenders for serious felonies, they issued non-judicial punishments (Article 15s, etc.) and had some reclassified to other USAF career fields. Pruning dead wood from the SP ranks made the unit stronger and eliminated the potential for corrupting influences. To fill the manpower shortfall, Colonel Allison requested 40 new cops straight from the SP Academy so he could have them trained and supervised correctly. He also got the USAF Personnel Command to rotate in seasoned NCOs whom he knew and trusted.

Several new K9 troops arrived when I did. A couple of months earlier, some young K9 handlers caught some intruders near the sewage lagoons on the north side of the base. They handcuffed and "dogged" them almost to death, reportedly leaving them for dead. The incident made international news, and Colonel Allison threw the book at the offenders. If I recall correctly, he sent one to prison, dishonorably discharged a couple more, and reclassified the rest to other career fields.

While my peers in training were soon assigned to shift work, I was directed (without explanation) to stay on for a third week. The trainers gave me a pile of manuals, advanced SP skill course books, and reports to read; they also had me take various exams, some of which were way above my qualification level. I was trained to drive the M706 "Duck" — a wheeled armored personnel carrier (APC) — and familiarized with the lone M113 tracked APC in the 3rd SPG section of the base motor pool. Both were easy to operate and reminded me of driving farm equipment. The nomenclature tag inside the Duck's cab identified its manufacturer as Cadillac. The M113 was loud and could only be driven on the gravel parking lot because the tracks were hard on asphalt pavement outside the motor pool.

Although it bugged me at the time, I later learned I was "picked out" for special duties. Why was simple: I could type (in that era, very few men could), was a crack shot with rifles and pistols, and had athletic ability in multiple sports. Many years later, Colonel Allison admitted his command team identified guys fitting my profile for special duties and assigned them to day shifts or administrative jobs so they could keep their athletic teams manned.

At the end of the third week in SPG Training, I was assigned to the Detention Center (aka the "Clark County Jail") and remained available for evening sports, honor guard duty, and special missions. This very greatly annoyed my FNG buddies, all of whom were assigned to swings and mid shifts. Half of the new LE personnel were assigned to the 7th SPS for 90-day stints to fill manpower vacancies around the flight line and at perimeter security posts. I took constant insults and grief over it. While everyone believed I received special treatment (in retrospect, I did), it was for reasons none of us understood at the time.

Detention was easy duty and somewhat boring. After day shifts the first week, I worked swings and mid shifts. Seeing a young guy working at the jail upset many mid-level NCOs who had long served their time on the line and were still stuck on rotating swing and mid shifts. On the other hand, the Detention NCOs (MSgt Dan Masola and SSgt Stovall) needed someone who could type reports and watch the prisoners. Although I never got to know Masola well, I liked Stovall, a Vietnam veteran and a base-level tennis champion.

The detainees were all US prisoners locked up for various crimes, either serious military offenses or felonies off base. Most were minimum-security detainees who signed out at 0700 each morning to work at their units and returned at 1700; someone from their units always signed them out and back in. I escorted the minimum-security guys to the chow hall two blocks away and had meals brought in for the occasional medium-security prisoner. In that case, either I or another jailer remained at the center for the latter. During chow, I kept an eye on the prisoners, and we ate in a separate area away from other personnel. I ensured that no one spoke to them and frisked them afterward for contraband (mainly eating utensils) that they could use to attempt escape. In reality, since most were white or Black Americans who would be easily recognized in a foreign land, the probability of successful escape was very low. Beyond that, those who had committed crimes off base did not want to be caught by Filipino police—much less stay in Philippine jails—so it was safer for them to remain in US custody.

There was a fenced enclosure out back, designed as an exercise yard for the detainees; in reality, it served as a smoking and relaxation area. Most new detainees spent two weeks on medium-security status before being downgraded to minimum security. Our only long-term medium detainee was John Krampf, a brilliant and intense man who had spent several years on international hold for drug smuggling. He completed at least one graduate degree by correspondence while locked up. The other detainees kept their distance from him, and he kept them in line for the NCOIC. Most prisoners were genuinely scared of him. John assured me that if anyone threatened or harassed me, he would deal with them. He eventually shifted to minimum security as his legal case was resolved, and he was eventually reassigned to the 3rd SPG. An electronics whiz, John lived in the barracks with us (along with his large pet bird) and installed duress alarms at all the gates. Just before my tour ended, he was discharged and went to work for a corporation in Switzerland.

One day, a unit brought in a large, heavy-set airman whose elevator obviously didn't go all the way to the top. He was incarcerated for writing bad checks, gross insubordination, and punching his supervisor. During his initial counseling, SSgt Stovall asked him why he continued to write bad checks; he replied in a hostile tone that it was because he had not run out of them. Stovall yelled, "Clark, come get this stupid bastard out of my office before I kill him!" I locked him up in the maximum-security cell and recall the young lad straightened up considerably after Krampf had — for lack of a better term — a "guidance session" with him.

I only heard of one detainee ever escaping, and it occurred a couple of years before my tour. Reportedly, he took off running while walking to the chow hall; he made it to the perimeter fence a half mile to the west and disappeared. A Filipino-American who spoke Tagalog, he was able to blend in with the locals. In 2000, while stationed in Arizona, I read in the Davis-Monthan AFB newspaper that he was finally apprehended in the Philippines and returned to US custody.

198401: Senior Airman Gary Bailey who worked Customs for the 3rd SPG in the Philippines. This photo (circa 1984) was taken at Davis-Monthan AFB in Arizona. He later became a world traveling model and graced everything from cereal boxes to hair care products to print ads. He now works in real estate in Florida. Gary, Matt Rossoni, and I still stay in close touch.

An interesting activity occurred while I worked at the jail. Concerned that the inmates weren't getting enough exercise, Krampf asked MSgt Masola if they could obtain some large boulders and a sledgehammer to create gravel. I thought giving them sledgehammers was a bad idea, but kept this to myself. Sure enough, it was approved. Soon afterward, the base engineers dropped a weekly boulder over the fence with a forklift. The inmates took turns beating the rocks into small pieces that they loaded into 5-gallon buckets for the base engineers to use for street projects. Krampf supervised this evening spectacle, and the inmates seemed to enjoy competing over who could make the most gravel.

After a month of jailer duty, an NCO from A Flight LE reported in, and I trained him during a couple of swing shifts. I was reassigned to Alpha Flight for patrol and gate guard duties; I was very pleased since I needed the experience. It also quieted the grumblings of my peers and the NCOs who perceived special treatment for an FNG who hadn't yet served his time on shift duty. My first supervisor, SSgt Fellows, was a real character and kept me laughing (although not to his face) with his antics. Although somewhat of an overbearing dickhead at times, he spoke his mind, and I learned all I could from him. Like many others, he had some bad habits, and I learned from those examples too, since I had no aspiration of emulating them.

Alpha Flight was led by a Philippine-American named LT Luena, who was an Air Force Academy graduate. He rarely spoke at guard mount and was almost inaudible when he did; thankfully, he let the senior NCOs run the show. Our Assistant Flight Chiefs rotated frequently, and all were good. My first one was SMSgt Turner, who later turned it over to MSgt Tom Loprete, a street-savvy New Yorker. CMSgt Harley Fields was the senior SP NCO in the 3rd SPG but occasionally served as our flight chief for a few days at a time; a highly respected and experienced NCO, he had served since the Korean War and was approaching mandatory retirement. Fields had bright blue eyes and would stare a hole through you when displeased.

My section leader was TSgt Garther "Cool Breeze" Halbert from Columbus, MS. Breeze, as we all called him, was an old brown-shoe who had served in Vietnam, Thailand, and previously in the Philippines. His wife was Filipino, and she owned a bar in Dau just outside the Main Gate called "Coolbreeze's 20-20," named after the cheap wine Mad Dog 20-20. It was a retiree hangout and most definitely not my preferred scene. When a couple of us visited the bar, Breeze wasn't there. A white-haired Black retiree drunkenly stammered, "The 20-20 is fo' real niggas only. All y'all white muf' uckas bes' get da fugg on outta heah." When we started laughing, the old guy picked up a bar stool and came after us. So we obliged and left, figuring discretion was wiser than valor. I told Halbert about it a few days later while he was dropping me off at the Main Gate; I mimicked the retiree's exact accent, and Breeze instantly knew his identity. Breeze had to pull over due to laughing so hard. Despite his invitations, I never visited the 20-20 again.

On one of my first days with Alpha, Breeze drove me around the perimeter fence and bomb dump in the Mabalacat Sector. He pointed out the holes in the fences and said thieves persisted in stealing large sections of it for sale on the black market. He said you could find the stolen barbed wire around the walls of homes and pig pens all over the province. Workers would quickly repair the perimeter and bomb dump fences, and thieves would soon steal sections again anywhere it wasn't under direct observation. They would sneak in, cut the wire brackets connecting the strands to the posts, then cut 1–2 sections out to lay flat. They rolled these up and shoved a bamboo pole into the center so two men could haul the section away toward the perimeter. Breeze asked me to take photos of multiple stretches of the newly installed fence and then take identical photos in a month for comparison. I did, and sure enough, the before and after photos showed significant chunks of wire missing. Within a month, 10-20% of the repaired bomb dump fence would get stolen and then repaired again, with lonely concrete posts as the forlorn indication that there was ever a fence there at all. The perimeter fence was even more bare and looked like Charlie Brown's Christmas tree for hundreds of meters in places.

Most Alpha LE NCOs were great to work with (among them SSgt Chuck Nadeau and SSgt Johnson). Married to Pinay women, they spoke Tagalog and taught us effective patrol techniques and things to look for that weren't in the textbooks. A few NCOs were average at best and marking time until retirement; these men enjoyed having the junior guys chauffeur them around the base while they badmouthed everyone else, politics, and life itself. Some wanted to park under large trees to "monitor traffic" — translated as "take naps while the junior guy monitors the radio." I avoided the latter NCOs to the extent possible since learning anything useful from them was a long shot. Honestly, if our roles were reversed, I would have driven the low-performing NCOs halfway out to Mabalacat Gate and beaten the living hell out of them for malingering, oxygen thievery, and dickheadery in general. I've always had a low tolerance for lazy employees, regardless of their rank or status.

Among the initiations for new first-term airmen was getting sent on wild goose chases to locate people or to find odd items. For guys on the flight line, it was often to obtain a bucket of prop wash or other fictitious maintenance-related items. For LEs, it was getting sent to the Main Gate bus stop to find some VIP, where at times during the morning rush hour, there might be hundreds of Pinoy waiting for pickup by base buses and unit shuttles. I had heard stories from older veterans, so I never fell for the stunts, but many did, and it was always harmless fun. One morning, as I worked the Main Gate pedestrian lane, I saw an LE patrol disgorge a young trooper who walked up to at least 150 Filipinos to yell out, "Colonel Ingus! Does anyone know Colonel Ingus? Colonel Ingus, are you here, sir?" The Filipinos snickered and laughed at the poor kid until he finally gave up and left. Another day, an FNG was looking for Mike Hunt and drew the same reaction from the many Pinoy standing in line. The goofy stuff never seemed to end.

Among my favorite mentors was a Georgian named SSgt Lewis Nunnally. He was a Vietnam veteran who looked and moved like he was much older than his actual age. He didn't talk much and on the surface acted aloof until you got to know him. Many younger guys didn't bother and openly made fun of him; a few insisted he had been around since Sherman marched to the sea or that he fought in World War I. Nunnally ignored them and went about doing his job. Conversely, I found him very wise and a pleasant NCO with whom to patrol. He was married to a Filipina of Chinese descent and had a little dog named Phoebe. Lewis was a hobby coin collector, and I bought a few silver MacArthur pesos from him to give to my dad. He stayed in the Philippines after he retired, and we remained in touch for many years until we lost contact.

Patrol duty was enjoyable for the most part. We didn't concern ourselves much with traffic enforcement or crash investigations since the 3rd LES had a motorcycle traffic section specifically detailed for that purpose. If we saw blatant violations, we issued citations — DD Form 1408 — but otherwise left this duty to the motorcycle cops. In addition to patrols for the cantonment area (main base) and housing areas, we had patrols designated for funds escort missions (Silver 1–3), two-man quick reaction patrols (Enforcer 1 and 2), two for the Wagner Highway, and numerous foot patrols for the base housing areas. The NCOs drew most Silver patrols due to seniority; depending on the day and funds collected, these sometimes concluded well before lunch or by early afternoon. Despite foot patrols being relieved of duty by 1400, I frequently drew them because many of my superiors disliked them. Everyone favored the Enforcer patrols due to their exemption from chow relief, which allowed them to drive around the cantonment area all day, waiting for incident calls. The duty uniform for Enforcer patrols and the three back gates (Mabalacat, Negrito, and Sapangbato) was fatigues, and many preferred wearing them to the Class B uniform, which consisted of trousers and a short-sleeved shirt.

In sharp contrast to patrol duty, gate duty was another story, depending on your gate assignment. The front gates (Main, Clarkview, and Friendship) were OK until it got hot; we wore Class Bs and worked outbound vehicles and pedestrians at them. Gate duty was long hours of standing around, waving through outbound cars, and checking IDs. We also conducted random vehicle checks for weapons, contraband, and excessive amounts of booze exiting the base. While we were to salute blue (officer) stickers on cars, most of us detested saluting officers' wives and kids. Some wives would physically stop if you didn't salute, and a few would then laugh when we did. Likewise, some officers' teenage kids would make faces, flip us off, or yell "f—k you" as they passed. This annoyed me, so I started taking notes and thereafter pulled over those vehicles for random checks. If I saw a piece of flaky stuff in an open ashtray that might pass for dope, I detained the driver and called for a K9 unit — actually got a bust from one or two. Other gate guards started doing the same.

What we didn't realize was that Colonel Allison had had his fill of gate guards complaining about verbal abuse from officers' wives and kids. Likewise, he was beyond tired of complaints from officers' wives bitching about his LE punks not saluting cars. So he officially changed the 3rd SPG policy to salute only vehicles with officers in uniform or occupants in civilian clothes who "reasonably appeared" to be officers. There were almost no "dependent husbands," so if a male was driving, you saluted and were right 99% of the time. For those who contested the policy and stopped to complain, a simple ID card check sufficed for a salute. Trash talk from a few female officers followed; however, the wing and base commanders agreed with Colonel Allison. We were grateful to him for changing the mindless policy of saluting blue stickers.

Friendship Gate was the first gate on Mitchell Highway heading toward the Main Gate complex. It was close to the Abacan River, and across the river bridge was Carmenville Subdivision, where many officers lived. There were no shade trees near this gate, so it was generally very hot duty. It was very noisy due to its close proximity to the runway's south end. On the aluminum roof support, someone had carved that an Airman Young was shot to death there in 1972. We were told about this incident during our SPG training — that two SPs, bored on a midnight shift, were playing quick draw, and one accidentally killed his partner.

Clarkview Gate was the middle gate and very close to the perimeter fence. It had tall trees near it that shaded the gate shack from the afternoon sun. The subdivisions outside this gate had a mix of officers and enlisted, plus a lot of SPs lived in the area. Along this part of Fields Avenue were a number of art galleries and craft shops, plus several hotels used by TDY personnel. The Clark Aero Club and its small runway were located diagonally north of the gate, and farther east in the distance was a small weapons storage area (WSA) widely suspected of containing tactical nuclear weapons. As expected, no Filipino or American general ever admitted to the latter, despite the multiple razor wire fences with sensors between posts, area lighting, the central alarm tower manned 24/7 by US Marines, and a "red wire" warning fence announcing deadly force authorization in multiple languages. We were always advised not to go near that WSA nor ask questions about it. Occasionally, SPs would be assigned there as extra security, as I would later discover during my tour.

Working at the Main Gate complex was decent duty. We could ask the sector patrols for latrine breaks and walk out the pedestrian lane to the adjacent Town Patrol building, where we can also get water. There were usually five or six PAF guards assigned, with two working the inbound lane on the gate and the rest (including an NCO) working the pedestrian lane. There was also a mini-BX and snack bar across the gate's parking lot, so food was usually available for the day shift and swings.

The PAF guards mostly spoke at least some English, and some were fluent, especially those from larger towns where English was a required high school subject. They often spent slow periods teaching us Tagalog or other dialects. This often led to mild fun as we tried out our new vocabulary on passing Pinoy, who—unlike the French—just laughed when we butchered words instead of correcting us. We learned all the slang and exclamations, too. For example, when a Filipino starts a sentence with "sus" (rhymes with goose), it means "gosh" in English; the more inflection or drawn-out one says the word, the greater the emphasis or shock. The same applied to hearing "aie" (sounds exactly as read in English), which just means "hey." "Oi" or "oy" meant "oh." One aspect I liked about Tagalog was that it had no diverse vowel pronunciations. You spoke the words exactly as they were written.

Some of the PAF guys loved to entertain everyone with their antics. One well-educated PAF airman named Garcia, from a prominent family, enlisted to avoid attending college. He enjoyed flirting with the women, and with them, he seemed quite the ladies' man. Upon seeing unattractive women approaching the gate, he liked to act as if he saw Medusa and pretend to slowly turn to stone. His NCOs tended to feign anger and yell at him, only to then laugh along and thus reinforce his behavior. One afternoon, as a US airman dropped off a rather ugly girl at the pedestrian gate, Garcia suddenly collapsed on the concrete, acting as if having a heart attack. It was so dramatic that I honestly wasn't sure and rushed to him, only for him to laugh and blurt out, "Sarge, her body's like Venus with a face like my dog!" I helped him up as the airman who dropped off the girl began cursing at him. Garcia snapped back, "Go take a shower, bakla! You stink!" The American looked at me to intercede over getting called a homosexual, but I backed up my PAF buddy. I advised the airman to not take it personally, that it was just a hot, boring day. Honestly, while these characters rarely insulted anyone openly, once backs were turned or the people were out of earshot, the silliness began, and there were no sacred cows. Everyone was fair game.

One morning, I nearly got jumped by a big senior NCO from the 374th TAW. A woman showed up at the gate without a pass, and we weren't allowed to let unescorted individuals out of the gate. The lady was nice about it and told me the man's identity; she said he was hungover and sleeping so hard that she couldn't wake him up. I called the LE desk sergeant, who in turn called the TAW orderly room, requesting that they wake the drunk and have him come to the Main Gate to properly check out his lady friend. Twenty minutes later, a very angry, large man wearing only boxers exited a taxi and began bellowing at me. He reeked of booze and said the lady was his friend, so I released her to leave the base.

When I turned around, he lurched toward me as if wanting to fight, and then heard a sound that froze him in his tracks. Actually, I froze too. The PAF sergeant had jammed his 45 against the man's left ear and cocked the hammer. We all three stood very still for a moment before the PAF sergeant calmly advised, "Big Sarge, you go home now. Or I make verrrrrr-ry big mess. You choose." By this time, the other PAF guards had their pistols out too and were aiming at him. This needed to end quickly and peacefully, or someone was going to get shot. I didn't want that, and sure didn't want to get splattered with this guy's blood and brains. The big American began shaking from fear, backed off slowly to the taxi, and got in before leaning out the window to yell and curse at us as the taxi pulled away. The two younger PAF guards waved their pistols and yelled for him to go "fornicate" himself too.

Meanwhile, MSgt Halbert had pulled up behind the taxi to monitor the situation. The PAF NCO holstered his pistol and was cool as a cucumber, as if nothing had happened. When Breeze walked up and asked if I was filing an incident report, I declined since the drunken NCO never actually touched me. He turned to the PAF sergeant and asked the same. He replied, "No, boss. You see big pussy-man shit down his leg?" I hadn't noticed in all the excitement, but there was indeed a trail of liquid poop starting a few feet away from us all the way to the curb. Breeze laughed and drove away. This story got around fast, and peers asked me about it for days.

Occasionally, early in the morning, Americans (some still drunk from the night before) would come through the pedestrian gate and give the PAF guards a hard time. This was typically over presenting ID cards or trying to bring in open containers of alcohol, expressly prohibited by regulations and base policy. One morning, just as I arrived at the gate, a shitfaced GI refused the PAF guard's order to pour out his drink and show his ID card. When I ordered him to comply, he yelled back "f—k you" and tried to push past the PAF guards; one grabbed him in an arm lock, and it was game on from there. As he got the better of the two younger PAF guys, their sergeant drew his pistol, and the American slapped it out of his hand. I leaped on the guy (he was about my size) and pounded his head several times with the butt of my nightstick, but it didn't seem to faze him. This frightened me since I was concerned he would grab one of our pistols and kill us all. With three of us on him, we finally got him down on the concrete. As we continued wrestling the drunk, I heard the PAF sergeant yell out, "MOVE!" I ducked aside right as he soccer-kicked the drunk under the chin. Now stunned, gagging, and bleeding from multiple places, his resistance waned as I finally got handcuffs on him. We dragged him over to a metal handrail and secured him to it with a second set of cuffs. If he had pulled the handrail out of the concrete wall, I'm certain the PAFs would have shot him, as they had already threatened to do. In that case, I would have stepped aside and let them.

A patrol came to retrieve the jackass and took him to the base hospital for a drug screening before taking him to our jail. We took turns going to the Town Patrol to wash off all the blood and tidy up. I was later told this airman tested positive for PCP, so that explained how he endured a thorough beatdown from four men before we finally got him under control.

Any time you walked out through the Main Gate pedestrian lane, half a dozen small boys were usually there to accost you for money, candy, or anything else you might be willing to give up. They would run alongside you, yelping and demanding, "Gimme peso, GI!" If you ignored them or at least apologized for not having any change, they quickly turned their attention to the next passerby. However, those who argued or reacted with even minor annoyance were treated to a recitation of very elegant combinations of cursing and invectives in both English and Tagalog. Some of these kids were 6-7 years old and could out-cuss most GIs. Other kids would try to sell you something—

gum, candy, Pepsis, or even one another for work.

"Buy gum! Buy gum! Buy gum!" "Pepsi! Pepsi! Pepsi!"

"GI, you want a nice girl? I know one—make a good wife! Love you long time!" "You numba 10! Give carabao short time! F—k you!"

"I shine your boots—10 pesos! Why not? Putang, GI!"

These were the common phrases I heard when walking through the pedestrian lane. I often wondered whether these street urchins were skipping school, feral orphans, or something in between. Obviously, begging worked, or they would tire of the activity and go elsewhere. It did work on me one time. Conversely, whenever I offered to buy these little buggers food, they always refused, as they really wanted money to buy things.

Once while patrolling the Mitchell Highway and Main Gate sector, an articulate teenager yelled at me through the perimeter fence trying to pimp out his younger brothers and friends to wash my patrol car. This set off an intense negotiation and we finally agreed on $2.50 USD (roughly 25 pesos). With no hoses available, I was curious as to how they planned to achieve this feat of creative cleaning. The boy asked me to stop by the gate in half an hour. I drove away and returned as promised to discover six boys with tabos (small plastic buckets) and towels standing under a shade tree just inside the gate.

The gate guards were also curious, so they let the boys enter.

The boys got busy scrubbing and wiping before I could get the car into park. They even wiped down the interior and cleaned the windows. The older boy barked orders, inspecting the work and directing the others to clean any smudges. Five minutes later, I had a spotless patrol cruiser. I paid the boys the $2.50 fee and tipped them two more quarters (30 pesos total). While this might not seem like much to most Americans, it made these kids feel accomplished and built their self-esteem. Sometimes thereafter, when I had time to wait, I drove out the gate and parked next to Town Patrol so the boys could wash my personal vehicle, plus I told my friends about these enterprising guys. I ascended to "Numba One GI" status with these kids. I took them party T-shirts from the Airman's Club, and they liked them even more than the tips. I sensed they'd never throw rocks at me and would willingly rat out any bad guys wanting to harm the guards at the Main Gate complex. All these little victories added up.

Although we wore jungle fatigues at the back gates (Sapangbato, Negrito, and Mabalacat), working them was downright awful at times. Whereas I didn't mind Mabalacat Gate because it sat directly under the shade of a giant tree, I somewhat disliked Sapangbato and Negrito Gates. Both were dusty with no shade, and the afternoon sun was always intense. During the monsoon, you got soaked and waded in ankle-deep water at times. The back gates reeked of urine because the gate guards routinely pissed behind the gate shacks or did so in bottles and poured them out nearby. One positive to working a back gate was getting assigned to a walking patrol or two during the same 6-day duty cycle. Most older airmen hated walking patrols, so I was often assigned two back gates, two walking patrols, one front gate, and one riding patrol during a cycle. Sometimes I received two patrols in one cycle, but not often.

Negrito Gate was named for the indigenous Ayta tribe that lived in a shantytown between it and the Bamban River. They picked up aluminum cans or pulled them from dumpsters and made souvenirs (samurai swords, knives, etc.) from them to sell back to GIs. They also flattened aluminum and tin cans and used them to shingle the rooftops of their shacks. They were resourceful and found creative uses for everyone else's garbage and cast-offs. The Negritos ("little black/brown people," named so by the Spanish) were mostly mountain people until WWII and rumored to have been cannibalistic headhunters. They certainly had no love for the Japanese and offered their services to General MacArthur. As a reward, he awarded them medical privileges for 30 years, and they lived on the base perimeter. Many worked for the 3rd SPG as DOD security guards and, for the most part, were dependable and honest. We treated them with respect, both for the latter and because their ability to infiltrate areas to exterminate anyone who crossed them was legendary.

I distinctly recall angering SMSgt Turner over letting a slop bucket get through Sapangbato Gate without a diligent check. Turner was going to formally reprimand me until MSgt Halbert interceded and got him to limit my punishment to a full week at that gate. Breeze stopped by Sapangbato Gate a couple of days later to see me. He asked if I had learned my lesson, and I replied that I had. He said, "That's my boy, because you're going to make us a good officer someday. Learn the hard lessons now so you can be a better leader then." He drove away before I could respond, but he obviously saw potential in me far beyond my role at that time.

Many gate guards (Americans as well as Philippine Air Force) treated the kids at the back gates with disdain or smacked them around, usually for climbing on the gates, yelling at the guards, and other minor annoyances. This always irritated me since it wasn't their fault for being born there; most were malnourished and skinny as rails. They weren't in school, so most were just bored. Many older people were aged far beyond their years, as evidenced by the yellowing of their eyes due to poor diets, and their life expectancy was much lower than ours. A few squatters and kids had lighter skin and blue or green eyes, evidently fathered by Americans who didn't know about or (worse) didn't care about having sired them. I felt deep empathy for these people and could do little more to improve their situation than treat them with kindness. Because I often brought little gifts (apples, candy) or gave the boys a few cents to sweep out the gate shacks, I generally got along well with the locals and kids around the back gates.

The worst part about Sapangbato and Negrito Gates was checking bike-mounted slop buckets for contraband like tableware, plates, etc. Locals would feed chow hall slop to their hogs, so taking a 5-gallon bucket home was a win-win for all involved; the farmer got free feed, and the base engineers didn't have to haul it to a landfill or unclog sewage pipes. Yet for us, having to hold your breath while waving off clouds of flies to stir a dipstick into what looked and smelled like vomit was sickening. There was no running water at those gates to clean up when you accidentally got slop on your hands or uniform, unless you used drinking water from the Igloo cooler. While I did my best, I must admit that I was rarely diligent in my searches.

The Filipino who managed the chow hall lived in Sapangbato, and he was a nice fellow. One day at lunch, they were serving steaks, and of course, the place was packed. Any SP wearing a firearm could skip to the front of the line, so the SSgt who picked me up and I went forward. When he saw me in line, the chow hall director brought me a huge T-bone, and the NCO with me asked why I got the bigger steak. Without batting an eye, the fellow replied, "Clark is number 1, Sarge." Of course, he sensed I would take half that steak to share with the kids at the gate — exactly what I did. Being nice to people costs nothing, and whatever reputation you built with the local Pinoy remained long-lasting.

The boys hanging around Sapangbato Gate constantly pestered us for money or anything else they could beg from us. There was a small sari-sari store 10 feet outside the gate, so I typically slipped out to buy candy or hopia (sugar bean) cookies for the kids early in my shifts there. Seriously, two quarters bought a lot of goodies, and I "rewarded" the kids when my shifts ended for alerting me to any vehicle or suspicious-looking individual approaching the gate from any direction. Just outside the gate was a row of two-story buildings, and behind them was the Abacan River. Although it was just a small stream during the dry season, it often became waist-deep during the monsoon and was subject to flash floods from heavy rains in the upstream mountains. The riverbed was the source for building materials and an endless supply of rocks for throwing. Unlike at Negrito Gate with its established sense of order, squatters around Sapangbato sometimes got into it with gate guards (Filipino and American) over perceived injustices. Occasionally, these amplified into brief rock barrages on the guards and gate shack. We always presumed that some rock fights were diversions to cover intruders and stolen property as they crossed the perimeter walls in that area.

During a midnight shift early in my tour, an LE got into it with Sapangbato's gate shack kids. Allegedly, he slapped one, and the kid ran off crying, only to return minutes later with a silver dollar–sized rock. The kid hit the airman square in the face and dazed him. Unwittingly, the LE drew his pistol and fired a warning shot — strictly forbidden by our standing ROE. This set off an hours-long rock bombardment that kept the airman and two PAF gate guards corralled inside the gate shack. At some point, the falling rocks tapered off, and the injured LE was taken to the hospital for stitches. I delivered the day shift LE guard to the gate the following morning and parked 50 yards away. I'll never forget the sight: hundreds of rocks scattered around the gate shack. I pulled Sapangbato Gate two days later, and things had calmed down by then, as if nothing had happened.

Someone in a building outside Sapangbato Gate had a stereo with big speakers. I never knew the identity of the owner or which building, since the speakers were out of sight. The speakers very often and loudly spat out endless rock music, thankfully accommodating my general music taste. Until that point, I had only heard the Scorpions (German rock band) a few times; however, I became very familiar with them in the spring of 1981. The mystery

DJ's favorite Scorpions song was "Always Somewhere," and one day it played over and over and over and over. If I close my eyes, I can still hear the chorus stanzas in my head and feel the late afternoon sun on my face as I waved away flies and smelled the vomit odor from slop buckets and scents from grilling meat riding the breeze from over the barbed wire atop the concrete wall.

Always somewhere

Miss you, where I've been I'll be back

To love you again.

Woo-oo-oo-oo-a-hoo.

This went on and on all day and was almost as annoying as hearing someone sing "Polly Wolly Doodle All Day." And I mean ALL the damned day ... ugh.

Once at Sapangbato Gate, a tall, older Filipino man with green eyes (obviously fathered by an American) stood mutely for over an hour, clutching the fence, staring down the road past the Armed Forces Network tower in the distance. He spoke perfect English and often talked to me about various topics; he told me he read everything he could get his hands on, so I occasionally took him news magazines. Anyway, I asked him why he stared so intently, and he replied, "I wish I could eat at the Chicken Coop." This was a chicken restaurant near the parade ground. The food there was OK — standard fried "change your digestive oil" fare — but nothing to write home about, so I asked him why. He replied, "To just taste it." Naively finding his response strange but thinking he might be serious, while on patrol a couple of weeks later, I brought him a small box of chicken. He thanked me without emotion, turned, and gave it to the boys hanging around at the gate. When I asked why he didn't eat it himself, he said, "I'd rather they experience it." Still too young to understand why he would do such a thing, I drove away in disgust. Another time during a boring day, I lamented the filth and piss stench around the gate to the same fellow. Without looking at me, he quietly replied, "Sir, you're lucky because you don't live here. Soon you'll go home to America and forget all about this place." I countered that I would remember my time here and all the things I learned. Still staring in the distance, he said, "It makes no difference." Taken aback but reflecting on it for hours afterward, I never complained about having to work the back gates again.

To do over now, I would sign that man into the base and take him to the Chicken Coop. I'd let him eat all the chicken he could hold and bring some back for those little boys. Despite seeing it firsthand, I nonetheless failed at this point in my life to fully comprehend the effects that abject poverty had on people. However, my experiences there taught me the value of frugality and to never waste food. To this day, I tend to order or dish out only what I'll eat and preserve the rest as leftovers. I saw enough starving kids in the Philippines and other places to internalize that most Americans are just one supply chain disruption from starvation themselves.

One afternoon, an old lady walked up to the Main Gate and asked what to do with a dead American expatriate. When I arrived at the armory from a back gate shift, I was "volun-told" to go help the Mortuary Affairs team pick up a deceased retiree's body off base. I didn't think much about it at first—just go along and learn from the experience. Three of us met up with an Alabama native named Mr. Mote and rode out in one of the mortuary trucks. When we arrived at the home, the decomposition odor hit us in the driveway. I wondered, How did the neighbors not notice this? Mote handed us rubber gloves and masks, then passed around a container of Vicks VapoRub to jam into our nostrils. "Y'all use this stuff liberally, and it'll reduce the smell. Trust me, I know."

We went inside, and the place was very stuffy and humid. The lady was his housekeeper and had been away for a couple of weeks visiting family. Mr. Mote guessed the man had died about a week prior, so with no air circulating, the body began decomposing fast; the stench inside the place was overwhelming, and the flies were awful too. Mote's team quickly sprayed the flies, sprayed air freshener, and opened the windows to air the place. As soon as I saw the body, my heart sank because this poor fellow appeared to weigh 300 pounds or more before the bloating. Yet Mote had a solution for this called visqueen plastic sheeting. "Boys, he's so advanced and swollen that we're just going to wrap him up, bed and all, and roll him out sideways through the doors." I was grateful the man died on a full-sized mattress and not a king or queen. We hoisted up each end of the mattress in turn to scoot it out from the walls, where we could wrap it up. The Pinoy mortuary guys wrapped up the dead man and his bed into a cocoon, then we lifted it up to put on caster plates. Just as Mr. Mote predicted, we rolled the entire package out the door and secured it in the truck outside. We covered it with a US flag and drove back to base. I'm not ashamed to admit we all vomited a few times during this grisly task. I felt sorry for the housekeeper, who had to clean up some soaked-through body fluids where the mattress had lain on the concrete floor. I never knew the man's name or whether he had any relatives in-country, just that he was a military retiree who died alone as an expatriate in a land far away from that of his birth. All I could tell that remained of his presence on our planet was an assortment of clothing and furniture, several rugs, a few books, kitchen utensils and dinnerware, and a set of dog tags on his bedside table. Seeing so little left me intensely sad. Once back at the barracks, I could still smell the terrible decomposition odor on me and taste it in my mouth, so I stepped into a hot shower without taking off my uniform or boots. I soaked for 15–20 minutes and had zero desire to eat evening chow. The next day, I arrived at the Armory thinking I was working a foot patrol and saw "EFD" (excused from duty) on the duty roster next to my name. The flight chief heard about the off-base detail and gave us three participants the day off. We were grateful for it.

One day, while on chow relief at the Main Gate pedestrian lane, a young Pinay with an older woman in tow came lurching up to the PAF guards. She was holding her pregnant belly and repeating, "Baby come! Baby come! Baby come!" Her water had broken, and she was definitely in labor. The older woman told the PAF guards that her daughter's husband was TDY to Korea, and they translated that back to me. I was trying to call the LE desk for an ambulance right as my LE patrol supervisor pulled up with the airman I had relieved for chow. I cannot remember the SSgt's name, only that he was from Louisiana; he ran up to the pregnant woman and shocked me by whipping up her dress. "The head's crowning, so this kid is coming now. Clark, let's go—you drive! I'm in the back with her. Might have to deliver this one."

We loaded the woman into the back, and her mom jumped up front, then off I sped toward the base hospital. I radioed the LE desk and reported the situation, so they had motorcycle cops stop cars at the big intersections. We arrived at the ER ramp, and the doctors took over, although the baby was already out. Our SSgt did a good job and shrugged off the doctor's telling him so. He said, "I grew up in the sticks and this ain't my first or even second delivery." He went inside and brought out towels to clean up the seat, then washed up inside before hopping back in. He turned to me and said, "Aight, boy, let's go get chow." Chow… not exactly what I was thinking about after all that excitement.

Around 0200 one Sunday morning that I was scheduled to have off, I awoke to loud banging on my door and voices yelling, "Recall!

Everyone, get up and report to the BCA!" I got into my jungle fatigues, grabbed my gear, filled my two canteens, and caught a ride in a passing jeep to the armory. Once there, the armorer issued my M-16, four ammo magazines, and a sealed plastic sleeve containing my special security instructions (SSI). About 40 of us—all new guys—clambered into the beds of two 2.5-ton trucks (better known as "deuce and a half," "deuce," or just "douche") for posting around the airfield as a direct security screen (DSS). Our truck headed north while the other deuce headed toward Barracks Row and the runway's south end.

We would man the DSS posts that stretched from the MAC Terminal to the east-northeast corner of the runway; from there, the second group would man the DSS line running down the east side of the runway down to the south end, then northwest to Barracks Row. That latter post was usually manned by an NCO, as it was near Central Security Control (CSC) and among the first relieved. It was a strange DSS because it was under a cedar tree at the V intersection of two main roads.

We rode in silence until encountering a K9 trooper who yelled for us to halt. He appeared from the darkness, brandishing his weapon, so the posting NCO stopped and spoke with him for a minute before moving on. We didn't know what the hell was going on other than sporadic chatter on our radios. The deuce dropped us in turn along the road a few hundred meters apart around the airfield. My DSS was the last post on the north end of the airfield. When we stopped, the posting NCO said, "Clark, your post is a sandbag bunker about 100 meters north of the road. Stay alert, exercise light and sound discipline, and read your SSI. We'll pick you up when this is over." With that, the deuce engine revved its whiny roar and drove away as the smell of diesel exhaust said hello to my sinuses.

It was pitch black except for ambient light from the cantonment area and the stars, so I wondered, How am I to read my SSI without turning on my flashlight? The sky was stunning like that at our farm back home and reminded me of a giant bluish-black blanket punctured by millions of tiny twinkling lights. I stood there for a few minutes to get my bearings and take a piss. Unsure what was up, I chambered a round in my rifle and turned the radio to its lowest volume. Once my vision adjusted to the darkness, I picked out the North Star and started off toward my post in the foot-high grass.

Before long, I found my "bunker" and was disappointed it wasn't one of the nice concrete pillboxes I had seen around the flightline. Instead, it was a circle of deteriorated sandbags on the grass that looked like a long-abandoned campfire pit. I thought, Shit, I guess they want me to dig a foxhole. I pulled out my entrenching tool and started digging. A few inches down, it bit into ancient volcanic ash, ideal because, despite its crunchy softness, it held up well for carving the sides. I dug for about 30 minutes, frequently looking up and listening for a few minutes in case someone approached. Despite the cool night, I perspired heavily. However, I now had a three-by-three hole with a short berm around the rim. When standing, my eyes were slightly above the grass, and no one could see me if I crouched down in a firing position. In the dim light, I could make out what resembled an uneven, fuzzy wall 50 or so meters north of my position — elephant grass. I had no idea what lay beyond it, but figured it was a big washout. I sat down in the hole and drank from my canteen, staring up at the stars. The radio was mostly quiet, so I figured this was just an exercise. Relaxing was a big mistake, for I soon fell sound asleep.

I was startled awake by a C-5 flying overhead. It obviously scared the daylights out of me because light now filled the sky. I looked at my watch, and it read 0620. My first thought was, God, someone could have sneaked up and killed me, followed by, Are they looking for me? After my heart stopped pounding, I checked my gear and had everything, so I started reading my SSI. I had read them before during training — just standard guidance. I figured I would sit tight until the posting truck returned. I was really upset with myself for falling asleep on my post. In the daylight, I saw that the dark wall north of me was indeed elephant grass. The airfield maintenance crews bush-hogged the airfield clear zone only a couple of hundred meters out and let the field approach beyond it grow wild. I emerged from my foxhole and walked up to the grass wall; it reminded me of a mature, dry cornfield back home.

Stepping into it a few meters, the grass dropped suddenly, and I stood on the cliff's edge of a huge washout — one of many expansive ravines that drained part of the airfield. Most areas were blackened from controlled burns and wildfires, yet a few areas appeared untouched. My post was next to one of them. Over time, monsoon downpours and erosion washed away the topsoil and volcanic ash to create washouts all over the base. We'd heard stories about finding unexploded ordnance, barbed wire, and other relics from WWII, plus everyone also knew thieves and insurgents used the washouts as infiltration points to the base. There were legendary stories of SPs getting ambushed and killed in the tall grass and washouts over the years, so I backed out and returned to the relative safety of my foxhole. I sat there thinking about all this, musing that posting us alone out there violated multiple tactical principles we learned at the SP Academy. I reckoned that the lowered insurgent threat in recent years made it an acceptable risk.

Around 0900, I was getting hungry, and it was hot with the sun beating down. I also had almost no water left in my canteen. Not having seen a posting truck and now hearing no exercise chatter on the radio, I suspected the exercise was over and I was missed. I hopped out of my foxhole and stood up to look around in the surrounding sea of grass. I grabbed my gear and walked back to the road. The perimeter was several hundred meters northeast, and the flightline was far off to the southwest. I considered walking down the road, but I rejected this notion and stayed put. Then it hit me: Call in for water. I called in a radio check to CSC. The voice replied, "You should have been picked up hours ago. Stay put. We'll send a patrol out."

Fifteen minutes later, a jeep pulled up, driven by a senior NCO whom I recognized as one of our many Vietnam vets. He stared at me quizzically and asked why I was so dirty. I replied that I had a bare fighting position and dug one for protection. He asked to see it, so we walked out to it. He jumped in, inspecting my work. He finally looked up at me and asked, "Is your weapon still charged?" I forgot that it was, but quickly admitted it, adding that I did so since I was posted without mission orders and didn't know what to expect. He stared at me for a moment as if in disbelief, then shrugged and said, "Good initiative. Unload your rifle and don't mention anything about charging it." We walked back to the jeep and a little while later, arrived at the armory. An older NCO in a passing jeep waved for us to stop, asking why I missed the pickup; the NCO who retrieved me replied that I had dug a nice foxhole and didn't hear the radio call. The inquiring NCO nodded his head and drove off. As I got out, the NCO warned, "Clark, I'm not gonna report you for falling asleep on post. But don't ever do it again 'cause next time it might be fatal. You gotta stay alert out there, boy." I nodded my head as he drove away. MSgt Breeze Halbert watched me turn in my M-16 and extra magazines at the armory window but said nothing. Fortunately, no one ever asked what happened, and I wasn't going to volunteer it.

Assigned to Chalk 40 for rapid deployment to South Korea in an emergency, I was sent to the ranges to qualify with the M60 machine gun and M79 grenade launcher. These weapons were respectively known as the Pig (due to its weight) and the Blooper (due to the sound it made). I scored expert with both and was identified for M60 duty with the Emergency Services Team (EST), the USAF version of a SWAT team. MSgt Funk insisted on having an expert gunner and picked me, presumably because I was also in top physical shape. The latter was needed since the M60 weighed 26 pounds, along with 500 rounds weighing 15 pounds more.

While I liked both weapons, I liked the M79 best. It worked like a single-shot shotgun of a much larger caliber — simple to maneuver with, aim, fire, and clean. It was versatile with any 40MM round: high explosive, illumination (white phosphorus), tear gas, buckshot, and the orange chalk practice round. The M79 was a great weapon, and I regretted its replacement a few years later by the M203 grenade launcher, albeit accepting the obvious flexibility gained from a rifle-launcher combo. Concurrently, I liked the Pig because it could pour a load of steel downrange, and any man-sized target I could see with the naked eye was in range of its firepower. The basic load of 1,500 rounds of ammunition (7.62mm or .308 caliber) weighed about 45 pounds, so the assistant gunner carried 2/3 of it. Additional ammunition was split up between squad members if on a combat foot patrol. Fortunately, those rare missions were exclusively for 3rd SPS guys who had attended Air Base Ground Defense school. Every exercise patrol where I had to carry the M60 was a mounted one.

We occasionally had rapid deployment force (RDF) alerts to support contingency operations in South Korea. I kept my RDF gear packed and ready in my wall locker. My A bag contained uniforms, underwear, cold-weather gear, field equipment, and a 30-day supply of toiletries. When an alert sounded, I grabbed it and headed to the armory to draw my weapon and gear (M60, helmet, and flak vest). Trucks then dropped us off at the MAC Terminal for the old "drag your bags" drill. We went through a processing line to check our personnel readiness, receive any due immunizations, and get issued any additional things we needed. We often loaded onto aircraft, not knowing whether it was the real thing or just a drill. If no ammo was on the plane for issue en route to Korea, we instantly knew it was only a drill. It's a sobering thought to see an ammo pallet on the C-130 and then someone break the security bands, since handing out ammunition follows. Once we flew to Taegu, Korea, and got out long enough to stretch our legs and use the latrine before reboarding and flying back to the Philippines. En route there, the plane flew through nasty turbulence over the South China Sea, causing a bunch of guys to barf lunch into their helmets. Fortunately, we were minutes from landing in Taegu when this occurred. From the air, South Korea was a beautiful country with endless green hills and valleys covered with rice fields.

Everyone knew there was an SR-71 Blackbird recon plane at Kadena AB in Okinawa, Japan. One day, as I was entering the chow hall, I noticed a bunch of airmen looking up and pointing at the sky. Far above the base was a huge white circle on the blue background, and people were jabbering on about what it was as we ate lunch. A weather specialist I knew was eating at an adjacent table, overheard us, and said it was an SR-71 at around 60,000 feet. He said the base weather station monitored this sort of thing. I never saw the Blackbird, only the contrail circle it left high over the base.

While I mostly ate in the chow hall, once in-country, I quickly developed a love for Filipino food. Some entrees, like pork adobo (braised pork cubes) and various forms of fried rice, are similar to Southern cooking. Pork sticks were basically pork kebabs. The best-known Pinoy food is probably lumpia — thin spring rolls, quick-fried to crunchy awesomeness. Pancit is a traditional noodle dish similar to fare found in most Asian restaurants; it's made in a variety of flavors and is very filling. If someone was making pancit, one could smell it beckoning from several blocks away.

Siopao is still my favorite Pinoy food. This is a big yeast bun stuffed with meats and veggies. Two would fill me up, and they were relatively inexpensive for their weight. One can find these in stores all over the world, particularly in places with high numbers of Filipino expatriates (the US, Saudi Arabia, Kuwait, UAE, Australia, etc.). However, if you eat siopao undercooked, you may get explosive diarrhea and experience the "Revenge of General Luna." I never developed a taste for the Kapampangan pork dish called sig-sig. It's a mix of "many pieces and parts" — noses, jowls, livers, dicks, and balls — that Americans tend to eat only in potted meat and beef sticks. To clarify reality, some hot dogs are made from lips, peckers, and other scraps not served in restaurants, so cringing at sig-sig is hypocritical. Anyway, for sig-sig, they added onions, peppers, curries, and additional items to the "meats" and stir-fried them. Overall, it tasted great but was way too spicy for my preference. Eating sig-sig required me to drink a six-pack of San Miguel's to flush out all the spices and salt.

Mongo is a sugar bean grown throughout the islands and a food staple. It's ground or chopped and added to a variety of foods, both for protein and as a sweetener. Mongo ice cream may not sound appetizing, yet I found it very good. Mongo hopia is a type of sugar cookie I ate regularly. A roll of six cost about 10 cents, so I bought it for a gate duty snack. There was an odd frozen treat called ice buko that was full of coconut flakes and mongo beans. I liked it, but rarely ate it for fear of getting an upset gut from ice made from untreated water.

Deemed a Pinoy cultural icon, balut is a partially developed duck or chicken egg. Filipinos love balut and often carry it to work to eat for a convenient breakfast, lunch, or both. Some GIs loved balut too, especially when splashed with a little hot sauce or curry. Whereas I disliked the quasi-rotting aroma, the sight of tiny feathers, beaks, and legs — at least to me — was absolutely heinous. Whereas my closest Pinoy colleagues insisted I was missing out, a few just frowned or shook their heads at my reluctance to even try it. To this day, I've never eaten a balut and doubt I ever will.

I did have a little fun with balut, though. As jeepneys full of Pinoy workers entered the gates, sometimes I announced a "balut check" to make the PAF guards and drivers laugh. Feigning annoyance, I demanded, "Nasaan ang balut ko?" — Where's my balut? Naturally, balut eggs appeared in a dozen hands, and occasionally, I received a worried look. I then smiled and said, "Mahusay — sige na!" Excellent — go ahead! This always seemed to draw laughter from all aboard. The PAF guards loved it, so I conducted balut checks for entertainment when traffic was light.

I really liked the squid crackers sold by the sari-sari store at Sapangbato Gate. They were in small one-ounce packs that cost a penny each. I typically bought a dollar's worth so the PAF guards, gate kids, and I could snack on them all day. The crackers had tiny reddish-pink chunks of ground squid cooked into them and were only lightly seasoned, so we never got thirsty from eating them. Sometimes, before a shift ended, I bought another bag of 100 to take back to the barracks. The lady who ran the little store liked my business and always had squid crackers in stock. It's a miracle I never got food poisoned eating stuff like that.

Bagoong (ba-go-ong) is a Pinoy garnish that I enjoyed in small doses. It's a fermented fish or shrimp paste with a very hot-peppery aftertaste, sort of like wasabi. The best way to eat it is touching a fork to it, then picking up the food to accompany it. Many Americans ignored the warnings and scooped a spoonful into their mouths, only to feel the stuff detonate in their sinuses and cause a rush of slobber, tears, and snot to flow forth. One time, as I watched several elderly locals eating balut with bagoong, one offered me a bite. A new guy who just arrived in-country took them up on it and almost dropped to the floor in agony. They grinned to themselves and nodded to me. One said, "Masarap, GI!" I thought, Oh yes, I'm sure it's good, but that poor FNG will never touch Filipino food again.

The Philippines is the land of plenty when it comes to fruit. We had mangoes, papayas, guavas, bananas, jackfruit, avocados, plums, soursop, durian, langsat, and many others. Of course, pineapples were grown in large fields and cost 20–25 cents each, albeit cheaper for locals. I enjoyed going to the small markets and picking up fruit to shove into a blender to make daiquiris. The more ripe the pineapple or bananas were, the sweeter the taste.

It was best to take a Filipino friend or three along when shopping to get good prices. If vendors saw an American, the prices were suddenly several times higher than for a Pinoy buying the same item. Close friends advised me, "Say nothing and let your local friends negotiate the price." This was sage counsel, since fixed prices were rare. Almost everything was negotiable. Similarly, if you wanted a local artist or craftsman to make something, you doubled the time for when you wanted it. In other words, if you needed something in 60 days, you ordered it for completion and delivery in 30 days; this allowed time for overruns, parts shortages, etc. The area around Friendship Gate featured many art galleries. I had one artist paint my high school graduation portrait for about $75 USD, and he delivered it in 30 days. Guy did awesome work. Americans based there and tourists from all over the Pacific region loved shopping in the Philippines. Craftsmen made beautiful capiz (shell) lamps that were very popular, as were carved wooden desk plates. The latter were typically 6" x 18" with one's name in the middle on raised letters; the borders were bas-relief depictions of palm trees, mountains, and carabao working rice fields. These might also have a base with pen holders and cutouts for staples and other office supplies. One could also buy a plethora of carved wooden items — chess sets, figurines, kitchen items (trays, utensils, sugar and flour containers), tribal ceremonial masks, etc. The best craftsmen could carve uniformed busts of officers that looked remarkably like those sculptors created centuries ago of Roman commanders.

Rattan and wicker furniture were available in every imaginable configuration. If you placed an order for a specific piece or style and allowed sufficient time, a furniture shop could make it for you. My great-aunt's son-in-law gave her a load of wicker from his Philippines tour in the late 1950s. As a kid, I always helped move my aunt's white-painted wicker furniture from her front porch into storage each fall. Each spring, I helped her clean and move it all back; every few years, we painted the wicker a brilliant white again. Anyway, many single guys who were leaving active duty typically got rid of uniforms and other items so they could ship custom-made wicker and rattan furniture home. The maximum weight limit for shipping household goods was based on one's rank. Because rattan was light, even a sergeant leaving active duty could ship a decent amount back to the States.

Probably the coolest things the Pinoy craftsmen made were the custom classic automobiles. There were multiple shops that could make these, and their work was not cheap. If you provided a full-sized car, a big down payment, and 12 months lead time, these mechanics and metal workers could make awesome replicas of Excaliburs, Mercedes-Benz SSKs, Rolls-Royce Phantoms, or other 1925–35 era classics. Using the chassis and engine as a base, they could transform a Chevy Caprice or Buick Electra into a fanciful classic that looked authentic from the outside. I knew a captain who had an Excalibur built that he shipped back to the US to sell; I recall him saying he could triple his investment on it. I recall another officer had a reproduction of Adolf Hitler's Mercedes 770 built that he also planned to ship home and sell.

One aspect of local shopping was somewhat alien to most Americans: many shops closed and activities stopped for 2-3 hours around lunchtime. Pinoy liked to break for long lunches, and some took naps after eating. Many got up at first light, worked until noonish, and then napped for a couple of hours during the heat of the day. It was common to drive by a dozen manual laborers sprawled out under a shade tree during lunch. In reality, this was common in the US—especially the Deep South—until the proliferation of air conditioning and the shift away from an agrarian society. I recall that when I was a boy, during the late spring and summer months, my grandfather routinely took long midday breaks from plowing fields to eat, cool off, and take short naps.

Three months after I began Alpha Flight, my patrol leader (Sgt Henderson, I think) wanted to apply for an opening on the Crime Prevention team. While this was rumored to be a great duty, no one knew exactly what it entailed. We stopped our jeep outside the Pass and ID building and went inside to meet SSgt Donald Quesnell, who immediately got to the point. "Can you type?" My patrol leader shook his head no. Pointing at me, he asked, "Can you?" I replied that I could and served as the first sergeant's clerk in basic training. "You look like an athlete. What sports do you play?" I told him football, track, softball, volleyball, and was learning to play soccer. "Can you hit what you shoot at?" I nodded, adding I scored expert with every small arm we had in the armory. He said, "OK. Dismissed. I'll be in touch if I want either of you." We left and returned to our patrol duties, figuring he wasn't interested in us.

Several days later, I reported to guardmount and had already drawn my pistol from the armorer before someone said my name was scratched from the duty roster. Beside it was scrawled REASSIGNED SPAC. I asked Breeze Halbert what it meant. He told me to turn in my pistol, and he would give me a ride over to my new assignment. He said little in the jeep before pulling in front of Pass and ID. "Make me proud for going to bat for you, boy. Now go see SSgt Quesnell." Turned out SSgt Quesnell had vetted several candidates with the 3rd SPG training NCOs and shift NCOs before selecting two. Breeze winked at me and drove away.

The other selectee, Airman John Granacker, arrived a few minutes later. We reported to SSgt Quesnell and he spent several hours detailing our duties and his vision for proactive law enforcement operations. His contention was simple: educating the public was a force multiplier for LE operations, and ours was a critical mission. He outlined the routine tasks such as Operation Identification (OPID), newcomer briefings, crime activity analyses, and residential security surveys (RSS) in both on-base and off-base housing. Don then unveiled a large program he was proposing to Colonel Allison: target-hardening the base housing areas. He promised that our work would be difficult but rewarding. Above all, he wanted our absolute commitment to his vision that he believed would ultimately save lives if local politics went sour. Granacker and I agreed, and another adventure began.

198105: Staff Sergeant Donald Quesnell conducting a residential security survey at a captain's home in May 1981.

While its responsibilities required more thinking and longer hours than LE shift duty, Crime Prevention (SPAC) was a much better duty than working gates. The biggest plus was working part of the time in an air-conditioned building out of the heat and monsoon rains.

We were co-located with the Pass and ID team, a collective of oddballs assigned there for "administrative" reasons—nearing retirement, pregnancy, convalescents on light duty, etc. The E6 supervisor was a fat, lazy, pretentious type whom I quickly grew to dislike. His subordinates called him "His Royal Highness" or "HRH," and he exalted over his little realm. He let his minions run an unauthorized snack bar, and it was rumored they skimmed the proceeds for God only knew what. While he was probably a decent SP at some point in his career, I sensed he was now just a burned-out dickhead marking time for retirement, so I kept my distance from him. One of his Pass and ID guys taught me how to laminate flies and other insects, which I gave to friends as goofy souvenirs.

In defense of the Pass and ID team, issuing passes, ID cards, and other registrations was a dull, thankless duty where the only advantage was working indoors. An unwed pregnant airman worked in another section and was as unfriendly as she was unattractive. She nastily yelled at me one morning when I told her name tag was on upside down. While it was indeed, this greatly offended her, so she tried to pull rank; I advised that I had more time in grade. She then reported me for declaring, "Rank among same grade peers is like virtue among whores." No admonishment followed, but thereafter I avoided her. Overall, they were a decent bunch despite their constant smoking and collectively heavier girth. That said, none possessed anything near SSgt Quesnell's intellect and mission-focused personality, so of course, some openly made fun of him and talked trash behind his back. I chalked it up to those poor souls suffering from SPSBS — "Small Penis / Smaller Brain Syndrome."

The first thing I realized was that since SPAC had no vehicle assigned, I needed a car. I couldn't afford taxi rides all over the base and to off-base housing areas; riding a bike in the heat (and rain once the monsoon started) was out of the question. So I sent home for money from my savings and bought a "Clarkmobile" — an 8-cylinder 1973 Dodge Charger. Clarkmobiles were cars shipped over years before and sold to successive owners as airmen transferred back stateside. Some cars were quite old, even dating to the 1950s. Filipino mechanics kept them running with spare parts, sort of the way Cubans have maintained American cars imported before the embargo.

1981006: This was my 1973 Dodge Charger in June 1981. While the big engine drank fuel, it handled well and accelerated like a scalded dog.

While my car looked good and ran well, it guzzled gas, and the concrete roads ate tires quickly. I soon found I wasn't happy with the car, but decided to make the best of it. Fortunately, I was allowed to maintain a mileage log for my official business driving and was reimbursed for it. I seem to recall I received 22 cents per mile, so a 20-mile round trip off-base and back yielded a $4.40 reimbursement. Assuming gas cost $1.80 per gallon at the base gas station and I burned two gallons for a 20-mile round trip, I "made" about 80 cents, not even enough to pay for tire wear and maintenance. Midway through my tour, I bought a new set of tires and cringed at the $280 price for purchase and mounting. While I lost money using my own car for business purposes, it was worth it for sheer convenience.

To digress slightly, there were some very cool Clarkmobiles around the base. Several F-4 Wild Weasel pilots had their cars painted in the camouflage "shark face" style to match their fighter planes. The pilots' names were stenciled on the driver-side doors, with their wives' and children's names on their respective doors. The best one was a mid-1960s Ford station wagon. I seem to recall that everyone with cars sporting these F-4 paint styles lived on base. It would violate security protocols and common sense to have one's name on a car if living off-base.

Driving in the Philippines was often an adventure because you shared the road with a variety of vehicle types. The larger vehicle

ALWAYS had the right of way due to sheer size (physics) and the clear mortal threat to drivers of smaller conveyances. Trains were the big kahuna. You didn't try your luck since trains won such collisions and created buzzard food from the losers. Semis, large trucks, and buses were next, along with loaded sugar cane trucks. Some sugar cane trucks on the way to market were dangerously overloaded and unstable. In a front-end collision, a motorcyclist was as likely to get suffocated by the load of cane cascading down as from the impact. Jeepneys, light trucks, and passenger vehicles were next. Motorcycles were in a category all their own since there were multiple types — touring bikes, dirt bikes, and "trikes." Darting in between other traffic, trikes were small motorcycles with sidecars added to serve as taxis; despite cheap fares, you had little protection in an accident. At the bottom of the driving adventure scrotum pole was bicycles and then pedestrians. If on foot, it was wise to cross roads quickly and not chance getting run over. Drivers sometimes sped up!

Drivers generally stayed in their lanes on highways, more so away from the big cities. However, the highway to Manila typically turned into three vehicles abreast going south by the time I reached the outskirts of the capital city. Almost every trip to Manila was like a scene from the Mad Max movies. As always, vehicle size mattered, and the big boys ruled the road.

Stray carabao often walked into roadways and sometimes even

along them. It was fairly common to round a corner and discover one standing still in the middle of the road. Once, while on the way to Subic Bay, I slammed on the brakes and black-marked the concrete when a carabao stood in the highway west of Dinalupihan. I screeched to a stop about a foot away, relieved I didn't hit it. I was less concerned with damaging my car than with having to buy that foolish animal. As if to insult me, it turned and butted my front quarter panel, scratching a spot with its large horn as it lumbered off the road.

Of particular note, speed limits off-base appeared just as suggestions. The few times I saw local police monitoring traffic, vehicles — especially motorcycles — were whizzing by them. It was almost impossible for them to chase determined bikers who turned down side streets. In all fairness to the cops, their departments could not afford good-quality police cars, much less interceptors. With multiple vehicles sometimes wrestling for position in the same lanes, radar was almost useless for the few police that had the equipment. As such, traffic police were typically relegated to monitoring traffic flow as a deterrent and investigating accidents.

The base exchanges at Clark and Subic Bay imported fast Japanese motorcycles for sale to Americans, many of whom later shipped them back to the US. These "crotch rockets" were relatively cheap and very popular. And as one might predict, they were the common denominator in many accidents that produced serious injuries and fatalities. Inexperienced riders and alcohol were frequently contributing factors.

Another aspect of motorcycles was their huge attractiveness to thieves. Once a stolen motorcycle went out a gate or hole in the perimeter fence, its probability of recovery was very low. One night, as a thief sped toward the perimeter fence at the south end of the runway, a DoD guard armed with an M16 stood in his path. When the thief gunned the bike as if to run him over, the guard shot the thief off the saddle, just like in an old Western movie. Either the late thief was a rookie or didn't know the guard was an Ayta. Play lethal games — win lethal prizes.

I quickly realized some peers working flight duty were resentful about my having yet another primo assignment. I dealt with it, yet was sensitive to the backbiting, which spawned several nasty encounters around the barracks. Some resentment was lessened because I was pulling honor guard duty that few others desired, and I frequently played sports in the evenings and on weekends; most guys felt their duty time was better spent drinking and carousing. Our commanders liked winning teams, and playing on one was favorable in any light. In reality, anyone with an axe to grind could have also interviewed for SPAC duty. My selection was incidental since I happened to ride in the same jeep as someone else interviewing for the job.

SSgt Quesnell and I soon discovered we were kindred spirits, and our work progressed well for a long time. He was a superb teacher and felt I was a willing disciple. I truly believed in our mission, despite the misgivings of so many others. I traveled with Don to several other installations around Luzon to perform crime prevention training to US and Philippine military personnel. The triad common to every criminal act is "means, motive, and opportunity." Whereas it is difficult to eliminate the means and motive, crime prevention focuses on eliminating, mitigating, or transferring the opportunity risk. The entire thesis of our operation was simple: it is easier to prevent crimes (proactive law enforcement) than respond to them (reactive law enforcement). Proactive law enforcement is easier, cheaper, and often costs nothing more than awareness and a little time; however, it does require educating the public and involving them in their own security posture. The challenge we faced? People are too often lazy, complacent, or disinterested to get involved, sometimes including the police force itself.

We often hear, "It'll never happen to me." So, educating the base personnel was a constant task. One of my initial duties was to market OPID to the base population. I scheduled appointments to engrave Social Security Numbers (SSN) on bicycles, motorcycles, stereos, and other highly theft-attractive property. Although the Privacy Act of 1973 prohibited using SSNs for routine identification, it was still standard military practice at that time. With concern over identity theft/fraud still years in the future, no one thought anything about putting SSNs on everything. I engraved stereos and bikes upon request, plus managed a few diamond-chipped engravers people could sign out at the LE desk. I obtained permission to hold "OPID Days" on Barracks Row and spent the better part of three weeks working on this initiative. If nothing else, the bad guys were taking note that items stolen from barracks were now identifiable. In retrospect, I sense most items "stolen" from barracks were simply sold to crooks or black marketeers, then reported as stolen to avoid accountability by the Merchandise Control Office teams.

Don taught me how to conduct a thorough RSS on military members' homes, including a few for Philippine military personnel stationed at Clark. These surveys were popular because they taught people ways to decrease their vulnerabilities to home burglaries and other criminal activities. I lost count of how many RSS I conducted (at times up to five per day), and they kept me busy both on and off the installation. We ended up "target hardening" two base housing areas by completing an RSS at every home, holding community "town halls" with the residents, and setting up neighborhood watch programs. As SSgt Quesnell predicted, the bad guys shifted their attention to adjacent areas.

Concurrently, we urged changing command policies to require protecting the personally identifiable information (PII) of military members and families. Large military installations produced a huge volume of used typewriter and computer paper waste (administrative and operations-related). Unless marked classified or confidential, units typically discarded used paper in common trash, and — albeit unthinkable today — back then, few took measures to protect PII like full names, dates of birth, SSNs, addresses, phone numbers, medical information, and information about military dependents. Many Filipinos recycled discarded paper from base sources to wrap food and many other uses, so it wasn't uncommon to buy a food item from a street vendor and spot names, ranks, and SSNs on the wrapper.

Don and I recognized this threat and badgered Colonel Allison and other leaders to change this policy. It came to a head when an alert LE gate guard seized a Master Mobility Roster going out a back gate. Had war broken out on the Korean Peninsula and several thousand troops deployed, that list would have given local criminals a fairly accurate idea which military homes were empty or had no adult male present. That did it for Colonel Allison, and he became increasingly vocal about protecting PII; he recommended an immediate new base requirement to shred all documents containing PII. Colonel Weiner endorsed the recommendation, as did Colonel Corder, so all units soon had to start separating such documents for shredding or burning. That said, it still took another 20-plus years to force the Defense Commissary Agency to stop requiring service members to write their SSNs and other PII on paper checks!

Regarding the latter, I submitted an Air Force Form 1000, Suggestion Form. Generally, if a submitter identified a vulnerability or cost-saving measure that a command could employ, the individual was eligible to receive a monetary award of up to 10% of the savings value. I submitted two such Form 1000s. The first addressed protecting PII vis-à-vis the Master Mobility Roster mentioned previously. As a reward, the base commander (Colonel Weiner) sent me a commendation letter and a $100 US Savings Bond.

I submitted the second Form 1000 after discovering the base engineers were not locking the water tower on Mactan Drive in senior officer housing. I reasoned that insurgents or common criminals could access water supplies and poison thousands of people, thus degrading operational readiness. Colonel Weiner sent me another commendation letter and a second $100 US Savings Bond. Although I was not going to receive a higher monetary award based on what "might" occur, submitting them was the right course of action for readiness reasons.

Don had a very old light blue Ford Rambler that a local garage owned by Mr. Medina managed to keep running. By no means a sports car, it was functional and low-key, unlike the muscle car I had bought. The first time I rode with him, I noticed a couple of cords extending into the dashboard, so I asked him what they were for. He explained the windshield wiper motor was burned out and he could not obtain a replacement, so he rigged the cords and pulled on them to wipe rain off the windshield. I thought, Holy moly, he's serious! When I rode with him again after the monsoon started, I discovered firsthand how the "passenger-operated wipers" worked. Don said, "OK, Dan-boy, start pulling the cords so I can see to drive. I don't want to have a wreck that kills both of us." I'm sure I looked like a weirdo yanking the cords back and forth while Don drove along in his little car.

Don lived in a modest apartment at the end of a street outside Friendship Gate. He had a cat whose name I don't recall and a dog named Phoebe. While Phoebe was already an old dog and shuffled along as if almost dead, she was super friendly. I seem to remember Don took her with him when he rotated back to CONUS in 1983.

Stationed in the Philippines for nearly six years and engaged to a Filipina (executive assistant to a high-ranking officer), Don was already well known by local Pinoy leaders, and his success made him increasingly popular. For the weekly Philippine Flyer (base newspaper), he authored "The Police Blotter." This was the crime prevention column filled with the latest criminal activities on and off base involving military members and dependents. An artist had drawn a very realistic rendition of Don answering the phone, asking, "You want it when?" This cartoon was the header for the column, so many people knew who he was at a glance. Using his razor-sharp wit, he skewered those who committed crimes and foolish acts alike. Even better was his weekly "No Class Award" that he bestowed upon some deserving jackass in hilarious prose. While working at the jail, the inmates loved reading and discussing the latest "No Class Award," plus they always hoped such notorious douchebags could join them as inmates. Like I said, Don was somewhat of a minor celebrity on base and especially within the 3rd SPG.

Conversely, the attention made Don equally unpopular with some of his NCO peers and a few higher-ups who viewed his crime prevention ideas as a waste of time and manpower. Quite a few crooks wanted to eliminate Don, too, and he received a number of death threats. He spoke Tagalog fairly well and knew the entire Angeles City area, so the threats never seemed to bother him. He lived off base among the locals and minded his own business despite his somewhat high profile. Conversely, the threats did bother me, and I did something illegal that could have gotten me in a lot of trouble. I bought my own pistol.

The economic wonderland called the Nepo Market ("open" spelled backwards) was a sprawling flea market where you could buy anything you wanted for the right price. I had become friends with several Philippine Constabulary police officers. One PC NCO I knew well enough to trust (I'll use the pseudonym of TSgt Santos) recommended I buy a pistol. He said eventually some communist or other "putang" might try to kill me while conducting RSS surveys alone off base, so I needed protection. When I argued that SPs were prohibited from owning personal weapons, he laughed and insisted no Philippine law enforcement agency or court would give a damn if I shot my way out of a bad situation. Beyond that, he reasoned, if I killed a criminal or NPA communist in a street fight, the US Government would have me out of the country in a matter of hours. His assessment was correct because I knew one American NCO who was put on a stateside-bound plane a few hours after a justified self-defense altercation resulted in a dead Pinoy criminal. Realizing a 3-man court-martial was better than getting carried by six pallbearers, I accepted his advice.

One bright Saturday morning, TSgt Santos and I met to eat a fine pandesal and jam breakfast at the bakery outside the Main Gate, then drove in his jeep to the Nepo Market. He took me to a licensed weapons dealer he trusted, and we haggled over a number of decent weapons. The 50-ish dealer was a pleasant fellow who spoke perfect English with no accent at all. He was accustomed to Americans and had no reservations selling to an SP like me. When I asked if he had many GI customers, he replied, "Oh yes. All the time." He had a few M1 Garands and carbines, sawed-off shotguns, M16s modified into carbines, and lots of pistols; he also had slap flares, grenades, and a lot of ammo for everything. While I wanted a newer pistol, they were too expensive, and the M1 carbine he really wanted to sell me wasn't an option. I settled for a beaten-up Colt M1911, a spare clip, and 32 rounds. It had US markings and low serial numbers, so very likely predated World War II. I had no way to know whether it was captured by the Japanese in 1942, then recaptured by US or Filipino forces in 1945. I test-fired my new .45 into a sandbag through a folded pillow to muffle the sound and confirm it worked. While the $250 I paid seemed outrageous, I was satisfied with the transaction. As a kid, I learned how to operate an M1911 owned by an uncle, so I was comfortable with using it. I had enough bullets to fill both magazines twice, excluding any I used for practice. Santos was happy to help me with the purchase, and the dealer gave him 200 pesos (about $20 USD) right in front of me for the referral. The dealer told me to come back anytime and bring more American customers; in fact, he mentioned a big discount if I brought him a bottle of Jack Daniel's whiskey. I thought, I'll come back another time with whiskey for some grenades. I realized this was obviously an "off-the-books" purchase, meaning we would trust one another to remain silent about it.

I had a spare tire in my car trunk and sometimes stashed my pistol underneath it. I already had slip knobs for the doors to deter breaking in with a wire coat hanger. My .45 primarily stayed in my briefcase's back folder sleeve, so the briefcase remained locked unless I needed something from it or went somewhere. When going off base, I placed it under a newspaper on the front seat. I kept my mouth shut about having the pistol and never told Don about it. I always felt bad about not telling him, yet sensed he knew. Despite his black belt in martial arts and knowledge of weaponry, I always figured he was armed too. Wise SPs kept their private armaments to themselves and didn't show them off. I also knew if I got in a real jam off base, my Dodge Charger was a 3,500-pound, 280-horsepower battering ram that ran like a scalded dog. Between my car and sidearm, I felt safe enough to go anywhere.

With my roommate on night duties, I traded rooms with Airman Robin Johnson (Goshen, Indiana), who was on Dennis's same work schedule. Dennis worked on a "Snake" (surveillance) team with SP Investigations, and Robin was an armorer at the BCA. I moved across the hall with Airman Dean Bidwell (Kendallville, Indiana). Dean and Robin were a couple of classes ahead of me at SP School, and I had gotten to know Dean fairly well. We liked the same music, and he introduced me to the sounds of Triumph and other northern and Canadian rock bands; in return, I introduced him to Southern rock like Lynyrd Skynyrd, 38 Special, and Molly Hatchet. It wasn't like we were headbangers or anything—just liked good rock music. He and I both bought ghetto blasters with headphones so we could listen to whatever we wanted without disturbing anyone else.

You get to know everyone around you in the barracks quite well due to close proximity. Although we had two-man rooms, the latrines and showers offered little privacy. We had a TV in the open dayroom that almost no one watched since we could only get Armed Forces Network and a few Filipino channels. Some of the latter were broadcast in English. People back home often poked fun at the drama in daytime game shows, yet those were calm compared to the Pinoy ones. People would dance around, break into song, and feign fainting and fall down over their winnings—great entertainment even if one didn't understand the words. There was only one phone in Barracks 7504, and it was in the upstairs dayroom across from the TV. I don't recall anyone ever using it, and it rarely worked; I suspect those living in nearby rooms kept it continuously sabotaged to prevent it from ringing while they slept. The only other phone we could use was in the Peacekeeper's Pub next door, and it usually worked.

198102: Barracks 7504 in Feb 1981

Right away, you noticed how many guys bought relatively inexpensive stereos and speakers that were way too big for two-man barracks rooms. The BX imported electronics directly from manufacturers in Japan, so troops could buy all sorts of things at wholesale prices with no sales tax. Mornings and early afternoons were generally quiet since those on swing and mid shifts were sleeping; however, by late afternoon, most were up, and the stereo dueling began. One could hear country, rock, disco, funk, and heavy metal thumping from every direction. Some guys had two Bose 901 speakers in their 12x12 barracks rooms — amusing since the NCO Club had only 10 of the same type of speakers in its massive ballroom!

Built around 1960, Barracks Row was roughly a mile long, paralleling the flightline complex running southwest to northeast. Individual barracks buildings were two-story concrete and divided into four bays, each with individual rooms; the latrines and showers were in the middle on each floor. There were 16 barracks along the row, laid out in diamond quads of four each, with a couple of buildings between each "V" facing the paralleling streets. Tennis or basketball courts and parking areas filled the centers of most quads. In our quad, the 1st Special Operations Squadron building was on the east side, and an AAFES building on the west. The AAFES building housed a small BX and a restaurant called Pizza Factory with a screened-in eating area with great ambiance. The food was good and priced right, plus it had a variety of cold beers on tap. It offered kegs of San Miguel beer for about $15 each (including the $5 refundable keg deposit); thus, this eatery had plenty of business.

198104: The Pizza Factory restaurant and a small base exchange were adjacent to my barracks. It had good food and we could buy kegs of beer there.

Barracks life was carefree compared to stateside bases. Every building had houseboys who took care of all cleaning and maintenance chores. They cleaned our rooms, made our beds, washed our clothes, shined our boots, and did all the mowing and landscape maintenance around the barracks. For these services, we each paid $27 per month—$7 as "dues" to the barracks manager and $20 to our houseboys. I always wondered where those "dues" went, but never asked. Considering the barracks maintenance and lawn mowing alone, it was a bargain. For the most part, our houseboys were honest and hardworking, plus the job was financially lucrative to them. Occasionally, things would go missing, and it was sort of tolerated unless a houseboy got greedy. I knew only one who got fired, and it was for stealing from several guys in a short time. I frequently tipped my houseboys (mostly with cigarettes doled out judiciously) and never had a problem. While I didn't smoke, I bought my monthly quota to give away as gifts and tips.

Eventually, Dean moved to a room downstairs, and I had a new roommate named Harps for a few weeks. He was married and lived in the barracks temporarily until his wife arrived. Dean and his roommate both moved to an apartment off base, so he soon approached me about moving into his room downstairs with him as my "ghost" roommate. While single airmen were not authorized to live off base, many did and paid for it out of pocket; however, they were still responsible for their barracks dues. For morale purposes, commanders "unofficially" allowed it since all the supervisors and flight chiefs knew where their subordinates lived. So I moved downstairs, paid Dean's barracks dues, and had no roommate for the last half of my tour. I felt the $54 per month I paid was steep, yet worth it to have a room to myself.

Most barracks quads had their own snack bar, and ours was the Peacekeeper's Pub. It was a low, almost flat-roofed building on the north side of the complex. It featured a long bar with stools, several eating booths, a small sitting area for watching TV and movies, a nice dart board area, and a back room with two pool tables and several pinball machines and video games. It offered food (instant and microwave varieties), ice cream bars, soft drinks for a quarter, and beer for fifty cents. It was a good place to unwind and very popular with guys living in the barracks ("barracks rats"). The main barkeeper was Jesse, a likable Filipino in his early fifties, and a couple of adult women assisted him at various times. Just outside the main door was a tailor shop run by a heavy-set Pinoy named Dick. Dick charged us a quarter per patch for our sewing needs and could make (or have made) almost anything you wanted. Many had him make "tour hats" with various things on them, ranging from the basic to the uncouth. I had Dick make several tiger stripe boonie hats for friends back home.

198103: Peacekeepers Pub in Mar 1981. Barracks 7505 was to right and Barracks 7504 is in background.

The big first-floor dayroom in Building 7505 next door was converted into a martial arts dojo. Our unit's instructor was a good-sized Pinoy named Ernie, a super nice fellow who had a constant stream of students. One I remember well was a reddish-haired young sergeant named Doug Hanson, who trained diligently before and after work. Another was Dan Hecht from Delta Flight LE. Both served long SP careers (Doug on active duty, Dan in the Air Force Reserve) before retiring as master sergeants. Both also continued their martial arts pursuits and eventually ended up in the Martial Arts Hall of Fame.

I sometimes played darts growing up and was fairly decent at it. In early April, I entered a dart contest and was surprised to get invited to play on the 3rd SPG team. The 3rd SPG admin officer, Captain George Bunuon, requested that I go with the team to Camp John Hay (CJH) in the mountains at Baguio to compete in the Kool Tournament. A couple of weeks later, I got on the bus with the rest of the team and headed north. The trip was a real treat. We crossed a seemingly endless stretch of rice and sugar cane fields before turning into the mountains up Kennon Road. I spent most of the winding climb through the mountains looking out the window at sheer drops—few with guardrails—into wild river chasms below.

Baguio reminded me of a much less developed Gatlinburg, Tennessee. Upon arriving at CJH, we stayed on base at the Irogot Lodge (named for the local tribe) and played darts for the next two days. Told the base was relatively safe, I went running each night before joining the other team members for a beer. Due to Baguio's elevation, the nights were still cool, so each evening we had a roaring fire in the main lodge fireplace. Although our darts team didn't place high, our bowling team won first place, and the golf team did well. There was plenty of free beer and food, so overall we had a great time. I did throw well enough to win several t-shirts and a carton of cigarettes. I traded the smokes for a bottle of whiskey, most of which the older NCOs and Captain Bunuon downed on the last night there.

One of the NCOs, a private pilot, had flown to CJH. He offered to let me fly back to Clark with him in exchange for splitting the fuel cost. Thinking this was an opportunity to see the country from above, I agreed. We took off to the east, and the airstrip ended abruptly over a drop-off of several hundred feet, so I was glad we got aloft with plenty to spare. I took photos along the way home, and my pilot friend was nice enough to let me take the yoke some.

After I bought my car, a few friends and I started taking road trips in the surrounding area beyond. While you stayed vigilant within 5 miles of the base and in big cities like Manila, you could relax a bit in the countryside. For starters, people didn't see Americans that often and were friendlier towards us than those near the base. Many older Pinoys were WWII veterans and often wanted to talk to us about the war or how Americans liberated them from the Japanese. Some would invite us into their homes, and others would offer us food and soft drinks. I'm convinced many would have given us the shirts right off their backs—good and honorable people, despite their comparative poverty. I took a lot of photos on my trips and wish I had taken more. The POW camp at Cabanatuan was a couple of hours away, so I drove there to see the place, albeit there wasn't much to see at the time other than a couple of monuments. I had lunch with a number of older fellows at a small café, and they talked for over an hour about the "Great Raid" that occurred there in January 1945. As a boy, my father took me to hear Major Bert Bank talk about his terrible POW experience at Cabanatuan and how the Army Rangers and Filipino Alamo Scouts rescued them.

One of the men insisted I meet the local provincial governor named Joson, the guerrilla captain who commanded the covering force that held off a Japanese counterattack. So, taking a leap of faith, I loaded up these friendly fellows into my car and we drove to Joson's office in a nearby town. His secretary ushered us in, and we filed into his office for a brief visit. I told Governor Joson I had heard Bert Bank speak, and his face lit up. Bank was sort of a celebrity after founding the Alabama Sports Radio Network, plus had served several terms in the Alabama legislature. I wasn't surprised that Governor Joson knew who he was and had even corresponded with him. We all shared coffee and talked before he needed to get back to work. He wished me luck in my career and invited me to come visit again. I drove the old fellows back to the café in Cabanatuan and headed back to Clark AB. I made it back to base running on fumes since my car drank so much gas. I have always regretted not returning to visit with Governor Joson or getting the base honor guard NCOIC to invite him to visit Clark for a monthly retreat at HQ, 13th Air Force. He struck me as a very nice fellow.

Something I learned quickly was the nuances of speaking to Pinoy, especially the younger ones. If I asked someone whether I could get to a given street or location, the answer was always "Oo-oo" — yes. However, I frequently discovered this wasn't necessarily so, for what they meant was "Yes, you can get there from here." After grumbling about this to an older Filipino friend, he suggested I revise my question to "Does this road pass by (the given location)," which would prompt the listener to tell me "no" and suggest the correct route. It took me a while to learn how to ask for directions, and once I did, I was rarely lost again.

I also learned that while Tagalog and Pampangan were similar dialects of the same root language, they were quite different, and one had to pronounce the vowels in every word. If I asked someone (especially an older person) a question in Tagalog, he or she might answer me in Pampangan, and this confused me. Pampangans would sometimes restate things in Pampangan too; for example, if I said, "Sus, mainit" ("Man, it's hot") in Tagalog, Pampangans often replied, "Oo-oo, mapali!" ("Yes, it is hot!").

We used to snicker over how the "p" and "f" pronunciations are swapped and sometimes merged in the Pampangan dialect. When I hopped into a barber's chair, a barber might ask, "Sarge, which way you want me to fart your hair?" Or a friend might say, "My pucking dog has pleas" or "Watch the big fuddles when you stef outside."

Some Philippine laws were unique and unfamiliar to Americans. The most significant of these were the "estafa" laws. These were basically laws regarding fraud and deceit, designed to protect citizens from promises or contracts made under false pretenses. Honestly, while I thought estafa seemed like a sensible law, like any other, it could be carried to extremes in some cases. For example, if a drunk man told a local woman he wanted to marry her in order to gain sexual favors, she could sue him under the estafa law to indeed marry her. Court cases involving estafa rarely ended in the defendant's favor (especially if there were witnesses or offspring involved) and typically were settled out of court for undisclosed financial sums. I've often wished US courts could impose estafa laws on political campaigns so we could jail elected officials for making campaign promises they never intended to keep.

The massive Subic Bay Naval Station was a couple of hours away and a favorite road trip destination. The route was southwest from Angeles City to San Fernando and then through the foothills of the Zambales Mountains to Olongapo. I drove down one time to see the USS Kitty Hawk aircraft carrier, and a couple of Sailors offered to take me and my buddies aboard for a quick tour. Looking over the side, it was a long drop to the water from the flight deck, so I sure wouldn't want to fall overboard. Having no desire to trek into the metal labyrinth inside the massive ship, we never left the flight deck. That was my "first, last, and only" visit to a US aircraft carrier in the Philippines.

198203: Road trip to Subic Bay Naval Station, March 1982. Clockwise from lower left: John Marino, Paul Weseloh, Greg Whalen, Dan Clark. Also a base track team member, Paul stayed for a career and eventually retired as a Chief Master Sergeant.

The Navy PX at Subic featured all sorts of unique items from all over the Pacific region. You could take the ferry across to Grande Island, home to the pre-WWII Fort Wint, and stay overnight for a nominal fee. The water was crystal clear, perfect for snorkeling and scuba diving, and loaded with fish, including very large sharks. I once had a close encounter with a shark twice my length; fortunately, it was more curious than hungry and left me alone after bumping my air tank. I stayed closer to shore and in shallower water in my subsequent trips to Grande Island. Of historical interest, Fort Wint's artillery batteries were destroyed by the US Army to deny their use by the Japanese as General MacArthur retreated to the Bataan Peninsula in 1942. I hiked along all the batteries and inspected the big guns still in place, taking note of the demolition gashes that rendered them useless.

Another road trip destination was Manila. We could split rooms in fine hotels for $20–30 each; most had good food, nice pools, and other great amenities. Walking along the Manila waterfront with its frequent spectacular sunsets was especially enjoyable. You could walk for hours and take in the nightlife at the many cafes and restaurants. Manila had a McDonald's restaurant, so friends would occasionally bring back Big Macs to base. Eating a cold or soggy McDonald's burger was always about novelty, not taste or nutrition.

That said, there were many desperately poor people living in parts of Manila. Passing by a large landfill, I could not help but notice hundreds of children digging through tall hills of trash, scouring for food and recyclable items they could sell or trade. I could only imagine the living hell they faced every day just trying to survive without sanitary conditions, proper nutrition, and clean water. While I was raised to appreciate and show gratitude for having the latter, seeing such sights reinforced why. As such, I've always detested hearing Americans whine about little inconveniences, as very few have any concept of how poor most people around the world really are compared to Americans. It also disgusts me to see people deliberately waste food, knowing there are others elsewhere who are hungry and could make entire meals from it.

Visiting the historic Corregidor Island, site of the final surrender of US and Philippine forces in 1942, was a highlight of one road trip to Manila. Except for vegetation growth and a few postwar buildings, it remained almost unchanged from its retaking by American paratroopers in 1945. Malinta Tunnel and the long, empty concrete skeleton of the once magnificent Topside Barracks were spooky. Not only did the Americans and Filipinos trapped on the island suffer immeasurably during the long bombardment following Bataan's fall in 1942, but so did the Japanese (most committed suicide rather than surrender) when US forces returned in 1945. I felt my skin crawl as if being gazed upon by unseen faces and eyes of restless spirits that demanded I understand the mortal horrors endured by the young men who fought and died there. While the island was something I had to see, it left me with a pervasive and unshakable sense of sadness. The decaying concrete of Topside Barracks reminded me of a great dinosaur's remains slowly being picked clean and worn down by the elements and time. I thought about this as I watched the island shrink in the distance on the return launch to Manila. The emotional effect on me was so strong that I never visited Corregidor again.

About 30 miles south of downtown Manila is 500-foot-deep Taal Lake, a picturesque caldera lake of roughly 154,000 acres with roads lining its rim. In the middle of this lake is Taal Volcano, an island containing its own small caldera lake and even a tiny lava peak inside. The latter looks strangely cool from the overlooks in Tagaytay City on the north rim. While Taal has burped out small eruptions every decade or so, in recorded history, it has never exploded full force like Mount Saint Helens, Pinatubo, or others of the past century. For decades, it has remained the smallest active volcano on the planet, yet it is among the world's potentially most dangerous super volcanoes.

A Filipino volcanologist I met at Clark explained why Taal Volcano is so dangerous: it's classified as a stratovolcano due to its size and potential threat. Should the volcano's core erupt, the cool lake water cascading down into the magma chamber below would explode, potentially setting off a major event at the high end of the Volcanic Explosivity Index (VEI). Each numerical increase on this scale represents exponentially higher danger. While Mount St. Helens (1980) was a VEI 5 and caused negligible global cooling, Mount Pinatubo (1991) was a VEI 6 eruption that decreased the average world temperature roughly 1 degree. Krakatoa (1883) in Indonesia was estimated as a VEI 6 eruption and created significant weather events.

Contrast these with Tambora (1815), which was estimated as a VEI 7 and set back global temperatures enough to cause worldwide crop failures and famines over the next two years.

Seriously? This absolutely occurred. Temperatures in North America dropped so much in 1815 that it snowed in Memphis, Tennessee, in June. Also, weeks before Tambora's eruption, General Andrew Jackson's men picked citrus fruit in New Orleans prior to that decisive battle in January. In that era, farmers planted orange groves as far north as Albany, Georgia. Tambora eliminated that. So while the experts and all the opinionated clueless argue whether human causes are creating climate change, decades of human activity can get dwarfed in one day by a single high VEI eruption. Our planet's last VEI 8 eruption was New Zealand's Taupo (roughly 26,000 years ago), and many scientists believe it extended the last ice age by many centuries. Suffice to say, billions of people could starve to death if we experience another VEI 8. And should our planet experience a VEI 9 event, I'll quote the Filipino volcanologist: "Humanity goes extinct in a matter of weeks."

Before my tour ended, I had traveled all over Luzon and to several other islands. I enjoyed my trips to Baguio, rafting at Pagsanjan Falls, the long beaches at Lingayen in La Union Province, and seeing the rice terraces at Banaue. High in the mountains, Baguio was the summer home of Philippine presidents and a tropical equivalent to Gatlinburg,

Tennessee (minus the tourist traps and traffic jams). I also trekked to see several volcanoes, including the breathtaking Mount Mayon with its perfect cone shape that silently "smoked" during my visit. I went fishing a few times off the coast and usually caught big fish that I gave to the boat captains.

Learning to surf on a longboard on the eastern coast of Luzon was among the highlights of my tour there. I surfed once at Baler Beach and several times at Dingalan — both in Aurora Province — and loved going, except for the long drives. Getting to either place meant driving to Cabanatuan City and then to Palayan; from there, I turned north for Baler or southeast for Dingalan. Although roughly 30 miles apart via straight-line distance, these small beach towns were about 100 miles apart by road. Separated by the steep and heavily forested Mingus Mountains and with no roads along the coast, driving between the two required a four-hour excursion, assuming traffic wasn't heavy, the roads weren't washed out, and there were no wrecks. Even then, I dreaded the constant threat of getting hit head-on or run off the narrow highways by a livestock truck, overloaded jeepney, Rabbit Bus, or produce delivery truck.

Departing at first light on a Saturday morning, I could arrive at either beach before noon and surf until dark; I would then get up early on Sunday, surf or fish until 1100, hit the road, and arrive back at Clark in time for evening chow. Late in my tour, once back on day shift LE duty, I could spend a second night if I wanted, due to having three straight days off. However, these road trips were expensive if I drove my car and went solo, less so if two more guys went with me. The latter were usually guys from the base honor guard or their friends who liked to surf. While I could likely make the beach trips on one tank of gas, I never risked running out and getting stranded, so Cabanatuan City or Palayan were good places to refuel and eat.

I surfed at Baler Beach only once because of the dizzying, frightening drive through the mountains, plus because of the latter, the drive was over an hour longer than to Dingalan. Baler Beach was the site of the surfing scene in the movie Apocalypse Now, so Americans tended to go there and not to Dingalan. However, Dingalan was a much easier drive due to the route along valley roads, plus if the surf was calm, nearby Matawe with its awesome rock formations was a first-rate place to fish and snorkel. Wherever we went, Pinoy surfers let us hang out with them and sleep overnight on the beaches. It was almost like something out of a movie, especially the time a full moon rose over Dingalan Bay. Not having a board was never an issue, as Filipino surfers always offered to let us borrow their longboards. As with other places far from cities and military bases, no one bothered us or our cars. Always friendly and talkative, the locals seemed happy to teach me and other Americans to surf. I always had cool stuff and T-shirts in my car trunk to give the surfers as gifts.

I mentioned the famous Philippine Rabbit Bus. This bus line started after WWII when its founders bought surplus US Army vehicles and rapidly expanded, hence its adopted "rabbit" logo. Its distinctive red buses seemed ubiquitous in Luzon. Several friends and I rode a Rabbit during a weekend trip to Manila, and it was a true cultural experience — no air conditioning, suffocating cigarette smoke, and every seat occupied by at least (emphasis on "at least") one individual. Across the aisle from me, an elderly lady clutched two large chickens with only their heads protruding from a denim bag. In front of her, a man held a shoat (young hog) in his arms; it wriggled and squealed as its nervous poop plopped into the aisle for other riders to track everywhere. The driver seemed to ride his air horn at everyone else — cars, trucks, other buses, jeepneys, Sasquatch, and even the Loch Ness monster. The return trip to Clark was equally exciting but healthier because I rode up front, where I escaped all the atmospheric nuisances except the driver's cigarette smoke.

While I always rode the USAF bus shuttle to Baguio, I knew a few who rode the Rabbits there. The route along Kennon Road had steep drop-offs down cliffs into river gorges — often with no guardrails — making the trip a bit harrowing even in a modern military bus. Those afraid of heights often screamed and gripped the seats hard enough to leave claw marks. I heard it was much worse in a Rabbit bus and for good reason: every year, at least one Rabbit would careen off the road and crash into the riverbed, sometimes killing or injuring passengers.

Always an adventure!

There was a cool monument just north of Mabalacat: the Kamikaze Memorial. It sat next to the highway at the edge of a giant sugar cane field. During WWII, the Japanese built a second large airfield here, and as US forces returned to the Philippines, the first Kamikaze squadron was formed at the Mabalacat East Airfield. A dark running joke among the Pinoy was that this was a "one mission airfield — planes took off and never returned." The memorial itself was rustic, a simple concrete block wall stating its purpose flanked by sections with painted text.

198108: This was the original Kamikaze Memorial north of Mabalacat in August 1981. It was at the site of the Mabalacat East Airfield where the first Japanese suicide pilots took off. When World War II ended in 1945, the farmers returned and their sugar cane fields swallowed up the airfield as if it had never existed.

One side told the airfield's story in English, and the other side was in Japanese. In the middle front was a small altar on which visitors could lay flowers and other mementos. If not for the memorial, one would never have known an airfield ever existed there, as its three-year presence was swallowed up by the sugar cane fields that returned immediately after the war.

East of Clark in the middle of the Luzon plain was Mount Arayat, a beautiful extinct volcano and the focal point or backdrop of countless photos and paintings. I took many photos of it and prized one I snapped of a brilliant sunrise behind it. Mount Arayat was the stronghold of the Hukbalahaps during their post-WWII insurgency, and we were always warned to stay away from it. Warnings aside, I knew numerous Americans who hiked on the mountain with Pinoy friends. I flew over and around it a few times with friends or during my pilot training.

198206: This is a gorgeous sunrise over Mount Arayat with a C-130 Hercules in the foreground. A former colleague heavily edited it for clarity and emailed it to me. Practically everyone who served at Clark Air Base took similar photos during their tours there.

Mount Pinatubo was a large peak in the Zambales Mountains to our west. I took a long day trip up it once with some Filipino police friends. We drove to a trailhead and hiked the rest of the way up. There were clear trails to get to the top, and the view was spectacular. By air miles, it was only 10 miles from our base, but the roads to the trailhead made for a long, bumpy ride. In retrospect, that trip was somewhat dangerous, but I felt safe with my armed buddies.

There was a huge petroleum storage (POL) farm on the southwest end of the base in the foothills of the Zambales range. Giant fuel tanks were built into the south side of a long ridge extending into the base.

North of the ridge, beyond a hilly open area, was a portion of the NCO housing, mostly large apartments. The hilly open area had a significant amount of barbed wire and tanglefoot left over from WWII. Prior to WWII, several US Army artillery units had positions in these hills. To my knowledge, the US 40th Infantry Division and Filipino scouts — including many Aytas — captured these hills and the long Bamban Hills ridgeline north of the base in 1945. I heard K9 cops talk about Japanese caves in the hills around the housing areas, but never saw any of them. However, there were several visible caves along the ridges within the POL farm; the outside faces of some still bore bullet pockmarks from the war.

Within the POL complex was the base's primary satellite communications station. Also known as SATCOM, it was a collection of small buildings adorned with white geodesic domes that housed receiver dishes. The complex had a high internal wall with a single gate leading into the POL farm that was manned at night and during alerts. I believe SATCOM was connected to the Regional Relay Center north of the Main Gate at Dau. These were signals intelligence facilities, and we left their operators alone, except to provide security and check IDs upon entry. On the ridgeline overlooking the POL farm and SATCOM to the south was an expanse of squatter huts south of the perimeter fence. These were sort of an extension to the barrio of Sapangbato. By looking at a base map, one could ascertain that some of the huts were built on base property. Years later, a group of squatters successfully won a court case to have the US government build them homes on the contested property.

During daylight hours, the POL and SATCOM complex was minimally manned by a couple of LE cops, but at night it was augmented by half a dozen more, including K9 teams. One night, a group of kids climbed over the perimeter fence and began throwing rocks at the guards. A young SP I knew named A1C Calderone was on the receiving end of the barrage. What the kids didn't know was that Calderone was a former baseball player with a golden arm. He picked up the rocks and threw them back, injuring several kids before they called it quits. However, it created an international mini-incident, resulting in the kids receiving treatment at the base hospital. Calderone was put on the "rubber gun list" during the ensuing investigation and received a minor reprimand. A week later, upon returning to duty, the armorer issued him a bag of rocks as a joke before issuing his pistol.

Shaped somewhat like a tadpole in the middle of the base was Lily Hill. The hill was strictly off-limits due to venomous snakes and unexploded WWII munitions. Overlooking the entire area with sweeping views for miles in each direction, it was a natural observation post and used as one both before and after the war. Like other hills of tactical value, the Japanese dug a honeycomb of caves into it and heavily fortified them. When American forces returned in

1945, they bombed and shelled the hill into submission. Surrender was not an option for Japanese infantrymen, and they fought to the last man or committed suicide. Legend had it that the Negritos (Aytas) helped retake Lily Hill. True or not, the Aytas did seize other area strongholds and reportedly had zero interest in taking prisoners. Allegedly, the Aytas sneaked into a field hospital and slit the throats of every other Japanese soldier for the demoralizing shock effect.

After the war, US Army engineers sealed the Lily Hill caves, but occasionally erosion revealed their locations. The caves were rumored to contain booby traps; even so, snakes were a far greater threat. Although prohibited, I took several "recon" trips onto the hill and found a cave I could get into, resulting in a later "spelunking" excursion. I did this out of curiosity and to find relics, but didn't find anything. Any surface finds were long ago hauled away by others going in for the same reason. I knew an LE NCO — Mark Old — who found a rusted Japanese Nambu pistol that graced the wall of his off-base home. I'm sure I could have found all sorts of relics with a metal detector. Ultimately, my trips to Lily Hill were covert, and I kept them to myself. I've since spoken with many other guys who have gone spelunking in the Lily Hill caves.

All sorts of interesting WWII artifacts surfaced around the base over the years. Both expended and live ordnance were buried everywhere. According to base engineers, after the war, a lot of stuff (including destroyed aircraft) was pushed into ditches and bomb craters, covered up, and soon forgotten. In addition to finds resulting from construction projects, ordnance frequently appeared in areas that had been washed out after heavy rains. While building the new post office next to Lily Hill, work halted while the EOD team removed several grenades and a mortar round. I recall another day responding to the golf course to guard a dozen or so artillery shells (uncovered by erosion from a big storm) until EOD could pick them up; although live rounds, no fuses were screwed in. Found ordnance was always presumed dangerous.

An older NCO, on his third tour at Clark, told me that his father had served there in the mid-to-late 1940s. His dad visited him during a previous tour and showed him areas out past the flightline where countless wrecked Japanese and US planes were buried in large trenches. He and a few others went out a few times during dry seasons to hunt these areas with metal detectors. He said in places they dug down only 1–2 feet and discovered intact wings and other aircraft parts. I always wondered why the US and Philippine governments didn't just melt down and recycle the allegedly hundreds of planes buried around the base. I presume it had to do with the sheer volume of wrecks and tangled metal left over from the wartime bombings.

Midway through my tour, a worker cutting grass near FM Hill found a skull. Skeletal remains were fairly common and typically

Japanese. I distinctly remember this one was Japanese since the responding Mortuary Affairs team discovered uniform buttons. They also found a corroded rifle, bayonet, and other gear near the remains. This man's family back in Japan waited in vain for their son to return, never knowing his fate but presuming he died honorably. While we were never told what Mortuary Affairs did with found Japanese remains, I always presumed they were turned over to the Japanese Embassy for repatriation to Japan.

While crossing the Pizza Factory's dirt parking lot one day, something caught my eye. A hard rain the night before had washed away a little soil. I leaned down and picked up a small-caliber slug. Looking around, I found several more of the same caliber, all with rifling marks from firing. I gave most to friends, and years later, determined these were 8mm pistol rounds, thus most likely fired by a Japanese Nambu. Because the parking lot was only 200 yards from the original flightline, it was easy to imagine a bored young Japanese pilot taking target practice at beer cans or a coconut tree during the occupation.

Late in my tour, I remember our goon squad (guys in trouble who were assigned shit jobs for the barracks NCO) found an intact clip of M1 ammo while digging a new barbecue pit next to the Peacekeeper's Pub. They asked First Sergeant Mohammed Johari if they could stop due to the potential for hitting explosives. He took the clip from them and ordered them to keep digging. Fortunately, they found nothing other than old nails and broken chunks of concrete. Before I transferred stateside, I was eating lunch at the chow hall when workers digging the new C-130 simulator building discovered a large bomb. The evacuation cordon stopped 100 feet short of the chow hall, so we were allowed to continue eating. The EOD team carefully extracted a 500-pound bomb, likely dropped by a US bomber before the 1945 invasion. They hauled it up to the Crow Valley Bombing Range and detonated it a few days later.

There wasn't much wildlife on the base. Occasionally you might see a feral cat but not often since Animal Control kept them in check. Some Filipinos considered dog (asong) meat a delicacy, so stray dogs were extremely rare. While driving around, I mused at all the crushed frogs dotting the roads. Several decades before, someone had the bright idea of importing large toads from another country (allegedly Australia) to control mosquitoes.

Everything I have ever heard or read about Australian wildlife suggests that most species are nocturnal, extremely resilient, and almost impervious to eradication at levels below the Four Horsemen of the Apocalypse. I was told the project failed, other than spawning a population explosion of unwanted frogs around the base and province. While we rarely saw the frogs in daylight other than fly-covered static displays on the pavement, at dusk they emerged and hopped around everywhere. They frequently hopped onto the roads and made a distinct "pop" sound when run over. When driving at night, I purposely squashed more than my share of them. A few airmen liked to walk around with golf clubs and "tee off" with the frogs.

Luzon Island is home to around 40 species of gecko lizards, and these nocturnal buggers were everywhere. Allegedly, when hungry, they emitted a hormone or chemical that attracted mosquitoes. They got into our barracks rooms, and our houseboys insisted it was good luck to have them "choose" your room, so I never bothered them. Other Pinoy were superstitious about geckos and would walk around a dead one rather than step over it; whenever I found a dead gecko, I would stick it on my car's dashboard to help ward off evil spirits and thieves. While most species were mute, the Tokay geckos were vocal, and their nightly calls sounded remarkably like "f—k you." I thought guys were kidding about these legendary reptiles until I heard one for myself. I never saw one, but I sometimes heard them in the trees around the base. I liked mimicking their calls and was quite giddy whenever they answered back. Conversing with lizards was most definitely an acquired skill.

Snakes were fairly common and ranged from bamboo vipers to cobras. The Animal Control team (several Filipinos supervised by an SP NCO) picked up cobras regularly from housing and other base areas. Because cobras were territorial and dangerous, the teams usually killed them rather than try to relocate them. The team members had their hides tanned for reupholstering cowboy boots. As such, I knew several NCOs who had really cool snake boots. My only encounter with a cobra was during a predawn exercise. I was on a security screen post east of the runway and kept hearing something bleat like a deer fawn.

It got closer in the darkness, and I called for backup. Central Security Control asked me to describe the sound, then told me to sit tight because help was en route. Minutes later, SSgt Felix with Animal Control pulled up in a pickup truck and asked me to point in the sound's direction. He turned his truck lights that way, and a cobra's head popped up less than 40 feet away. He waved his flashlight and yelled at it while his Filipino colleague got behind it with a snare stick. Within 30 seconds, the cobra was in a tied-off burlap sack. Felix asked if I had charged my rifle in all the excitement; I nodded that I had and asked if he was going to report me. He grinned and replied, "Hell no. I would have done the same thing, kiddo."

A few months later, I saw Felix at the NCO Club, and he asked if I had seen any more cobras. I told him that thankfully I had not and didn't want to either. He pointed down at his snake-hide boots and said, "Here's the one you called in that time." He explained he had been looking for one to match a similar hide he already had to cover a pair of cowboy boots. Felix bought me a couple of rounds as a finder's fee. Lions and tigers and bears — and really big snakes too!

On the topic of wildlife, the K9 section was the wildest bunch of humans with whom I served during my entire career. The K9 barracks were in a complex on the south side of Lily Hill that also housed the US Navy contingent, Hospital Squadron, and Billeting Office, plus had several transient (TDY) barracks. This complex was adjacent to the Airman's Club, so the K9 guys frequently "prepped" there before trekking downtown. Their barracks had a snack bar that was intermittently closed for various reasons — lack of Pinoy brave and tolerant enough to work there among them. The 3rd SPS had a hard time keeping furniture in the snack bar and the barracks in general because the K9 guys had a penchant for "soiling" (use your imagination) and wrecking furniture. This wasn't the inexpensive, snap-together stuff one finds today at Walmart, Target, or IKEA. In that era, the furniture was heavy oak and corduroy fabric variety made by the US Bureau of Prisons and Industries for the Blind. Instead of submitting work orders to clean or repair items, the K9 guys periodically piled it up out back and set it ablaze, dancing around the pyre in various stages of dress (if any), headhunter masks, and green face paint.

Naturally, no one ever saw anything.

Getting invited to a K9 party was a treat, and skipping one risked never getting invited to another. The invitation was verbal or announced by something nasty shoved under your car's wiper that read "URINE-VITED" with a date and time group. They never stated a location since this was understood. It was unwise to park out front since that meant running a gauntlet of empty San Miguel bottles, rotten fruit, dog shit, or much worse (again, use your imagination) flung from the second-floor balcony or roof. I always parked in front of the Hospital barracks and walked in via the side door. The general theme of the K9 parties was the same as that of some college fraternities: No party really gets underway until the first piece of furniture breaks, a VIP is thrown in the pool or golden showered, or the host's dog bites someone.

My introduction to the first one was memorable. Once inside and safe from thrown objects, a sergeant wearing only a jockstrap and tiger stripe hat greeted me, then ushered me out back toward the barbecue grills they "liberated" from the Navy barracks next door. I was surprised at how many half-dressed people — both genders, plus a couple I wasn't sure about — were standing around eating, drinking, and talking. One guy held a giant lizard, and another had a small rodent, obviously trying to coax the lizard into eating it. Still another guy was wearing only flip-flops, a small towel, and a big snake around his neck. My guide said, "Steak's out here, sex is in Bill and Mike's room, plus more upstairs, and we got plenty of beer — three kegs.

Liquor is everywhere. Help yourself." I thanked him as a guy named Carl handed me a partially drained bottle of gin and said, "When you need to piss, make sure you piss in the big plastic jugs. Two guys got letters of reprimand from the first sergeant, so they want to water his flowers and front yard with it. OK?" I handed back the gin bottle without drinking any and thought, Wow!

Throughout my 18-month tour, the K9 guys continued to throw crazy parties with endless cold beer, grilled shrimp, and steaks served on big knives, Ayta-made samurai swords, or bayonets. I saw hilarious "dog races" behind the barracks, with hooker-jockeys beating their male mounts with rolled-up patrol leashes to make them crawl faster. Scantily clad or naked hookers paraded equally naked dog handlers around the halls on their K9s' choke chains, occasionally beating them with the leashes hard enough to leave welts and bruises. Seeing guys hike their legs and pee on things was disturbing, yet not at all surprising. At various intervals, they all lined up for a "guardmount inspection" that featured shocking penalties for infractions — anything to entertain regular guests and first-timers. These parties were pure hedonistic debauchery that would have made Emperor Caligula feel at home. As messed up as they were, I did my best to make every K9 party to which I was invited.

The "K9 Staff Car" was a military oddity of sorts. It was an old mid-1960s muscle car that someone brought over and sold to a friend, so over time it became known as a Clarkmobile. It was jointly owned by several K9 troops who successively sold their "shares" upon rotating out to new guys transferring in. It was painted battleship gray, except for an ancient and cracked Confederate battle flag on top. A large white star was painted on the hood and each front door, along with the inscription "Clark County Sheriff (K9)" to identify its owners. They jerry-rigged the trunk lid for easy removal to jam in a third row of seats with seat belts in the trunk area; this allowed them to legally squeeze in 9 guys for trips to the chow hall. In their words, it was "a classic hybrid of Redneck and Filipino automotive engineering," yet it was quite functional. In reality, and due to low speed limits on base, they frequently crammed 10–12 guys in the car. If the brake or turn lights stopped working, the driver yelled out "BRAKE" or "RIGHT / LEFT TURN," and everyone else echoed him. This signaled the guys riding in the back to use their flashlights to signal stops or turns. The Traffic Section's motorcycle cops ignored the K9 Staff Car, as did the LE patrols.

It was harmless fun and tolerated, as no one really cared if the boys tore up their own stuff. Every so often, the K9 guys caught big rats and superglued them belly-down on the staff car hood. Whenever they had "live hood ornaments," they parked in the shade and fed their pets to keep them alive as long as possible. When the rats inevitably succumbed to heat or dehydration, the guys just let the sun bake them into hard, crusty shells. When someone caught a replacement rat, they scraped the dead one off to make room for the new honoree. During my tour, I saw several live hood ornaments and felt this activity — albeit sick — was rather unique.

Over the years, whenever I shared stories about the K9 guys in the Philippines, listeners either refused to believe me or asked, "What made young Americans act like such wild animals?" Having decades to reflect and read other historic examples from Greek hoplites in the BCE centuries to German U-boat crews during World War II, I can condense my answer to this:

They were young people far from home doing shitty, dangerous, and often thankless jobs. Our K9 teams worked primarily at night, chasing after people who would kill them and eat their working dogs. Frequently, in near darkness, they patrolled on foot in sometimes difficult terrain while being bitten by mosquitoes and other venomous threats. They sweated in the heat or shivered from cold rains, often one footstep away from imminent danger (crippling wounds or death) created by natural or human threats for which they used their combined senses to detect and overcome. The K9 troops and their Military Working Dogs were alone out there in the dark and had to depend on one another and other friendly forces. While help was supposed to be only a radio call away, fickle batteries occasionally failed, leaving these kids and their dogs incommunicado. Firing a flare to signal for help ruined their and their MWDs' night vision; thus, they had to make fast judgment calls on whether to stay put and engage or break contact and maneuver to more defensible ground.

While any type of law enforcement or security duty is stressful on a scale, some duties are particularly miserable and draining, both physically and mentally, as well as psychologically. The K9 duty in the Philippines fell into the higher end of that stress scale, surpassed only by supporting infantry in combat, as some of the older NCOs did in Vietnam. As such, the younger K9 guys immersed themselves in their duties, worked hard, and played even harder. According to my K9 friends, it built such camaraderie that they wouldn't have traded dog handling for any other duty in the 3rd SPG. And without indicting or excusing them for their behavior choices, the preceding should help illustrate why they acted the way they did.

Shortly before my SPAC duty, I started dating another service member who was absolutely wrong for me in every way. Although a solid performer on duty from all accounts, she had very low self-esteem, smoked excessively, and drank like a Viking. Trying to help her disengage from her bad habits made things worse, and after a couple of months, I broke up with her, unwilling to get dragged further into her worsening personal problems. Presumably to get back at me, she slept with an NCO in another unit who passed on a penicillin-resistant social disease to her. During her treatment, she reported me as the source — unlikely since we had split up several weeks before. I got a call from First Sergeant Johari to go get checked out at the hospital. Given a clean bill of health and falsely accused, I was exonerated. I heard later that this woman was issued an Article 15 (severe reprimand) for falsifying an official report. From that point on, I avoided contact with her outside of official business.

Not long after that breakup, I started dating another airman who worked at the base hospital. It never grew serious, it became more of a platonic friendship than anything else. We hung out for a while, but I was too busy with work and playing ball to suit her tastes. However, she did introduce me to several friends with whom I remained close at my future assignment in Texas.

Interestingly, she soon met an airman who was in the hospital for surgery, and after a whirlwind courtship, they married. We reconnected via social media over 40 years later, and I was happy to discover they were still together and enjoying their grandchildren.

My only other relationship (late in my tour) was with an officer's daughter who was a senior in high school. We were only a year apart in age, so age wasn't a legal issue. Although cordial to me, her family openly disapproved, and this made things difficult. They wanted nothing to distract her from attending college immediately after graduation, and after a while, I purposely began withdrawing to acquiesce to their wishes. While I felt bad about it at the time, in retrospect, it served a higher purpose. Both young and bound for diverging paths in life, neither of us needed to embark on anything serious.

In late June, SSgt Quesnell informed me that I was a finalist for 3rd SPG Airman of the Quarter. Not having prior knowledge of my nomination, I was surprised and directed to study a binder full of potential questions that a panel might ask. On the interview day, I reported to the 3rd SPG HQ with other nominees. The panel consisted of LtCol Quist, CMSgt Fields, SMSgt Fred Carrender, the 3rd SPS CMSgt (whose name I've forgotten), and a board recorder. I knew the answer to every question and felt it went well. An hour after I returned to work, the phone rang; SSgt Quesnell told me, "Dan, you won." The next day, I had my photo taken at the base photo lab, and it was soon posted on the wall at the 3rd SPG HQ in a section dedicated to the NCOs and airmen who had received these awards. My peers jokingly called it the "Ass Kisser Wall of Shame." Colonel Allison mailed a letter to my parents to inform them of the award. He presented me with a congratulatory memo, along with a 4-day pass and a weekend of free lodging at Camp John Hay. Incidentally, I stayed so busy that I never took the pass or the free CJH lodging.

198107: Photo for Airman of the Quarter (April-June) taken in July 1981. I was very thin from running on the base track team.

A week later, I interviewed for the base (3rd CSG) Airman of the Quarter. I passed the uniform inspection and close order drill routines, but missed a single leadership question during the board interview. I immediately sensed that missing it cost me first place, and learned later that I indeed placed second. Along with other decent prizes, the first-place winner was awarded a back-seat flight in an F-4 Phantom fighter. As the runner-up, I received a free specialty dinner for two at the Airman's Club and treated a friend to a great meal.

In early July, Colonel Allison hosted the 3rd SPG commander's call at the Kelly Theater. There were several briefings by various staff members before our commander awarded the quarterly awards. He started with the DOD civilian award, then the NCO of the Quarter award, and finally called me up. I saluted and received my award, then Colonel Allison asked me to wait for a moment. He announced to the crowd that he anticipated I would get out in three years to attend college and then return as a Security Police officer. "As such, I want you NCOs to train this one right and get him ready for higher leadership roles."

Yikes, he surprised me with this, and I could see NCOs in the crowd groaning and making obscene gestures. A few were going to bust my chops at every opportunity until I rotated back to CONUS in a year; they were going to get in their punches now in case Allison's prophecy became reality. However, most NCOs did the opposite and took the time to share their collective wisdom, quiz me on problem-solving scenarios, and conduct other developmental counseling sessions. Coolbreeze Halbert, Sgt Mark Old, SSgt Johnson, SSgt Lewis Nunnally, SSgt Chuck Ladeau, SSgt Chuck Duvall, TSgt Tom Loprete, MSgt Don Funk and a number of other NCOs were first-rate trainers, team-oriented, and treated every subordinate with respect. And we junior guys loved them for it.

By July, Granacker quit SPAC and returned to LE flight duty. He was replaced by Timmy Tinker, a married A1C with a wife of the same age. While we got along at first, both of us were somewhat high-strung. While I felt no sense of competitiveness in our peer relationship, he obviously did and insisted on rubbing in mistakes — no matter how insignificant — that I or anyone else made. Beyond the annoyance, it was petty, abrasive, and distracting to our mission.

Don asked me to serve as a groomsman in his wedding in August. I was honored to do so and agreed. This was a true cultural experience for me, as it was the first Philippine Roman Catholic wedding I ever attended. I had a tailored tuxedo made a bit large to ensure I could still wear it for several more years. (I wore it to my 5-year high school class reunion and was WAY overdressed; then the final time I could fit into it was for a fraternity formal in 1987.) Belle had several friends as her bridesmaids. All were some years older than me, so I didn't talk to them much; one told me I had a "strange accent," and I found that somewhat amusing. I had never attended a wedding where doves were released, so that was rather cool. I most definitely enjoyed the food. Lawd help me, Pinoy food was so good!

In retrospect, I sure wish I had gotten to know Mr. Maico (Belle's father) better. He was captured on the Bataan Peninsula in 1942 and survived the Bataan Death March to

Cabanatuan. After a while, the Japanese attempted to show goodwill by releasing thousands of Filipinos from the prison camp; doing so backfired since many joined the resistance instead. Mr. Maico and others served in the resistance until liberation, and he was a true hero to his people. I didn't appreciate the degree of his wartime service at the time and regret not recording his stories. I sensed he didn't like talking about the war, so I let it go at that. He and his wife came to Texas many years later to live with Belle and Don. Mr. Maico finally returned to the Philippines in the 1990s and lived out his remaining years there.

While away for his honeymoon most of August, Don assigned me to write several editions of the "Police Blotter" for the base newspaper. I submitted all four articles to the Philippine Flyer staff on time, and other than mentioning one sensitive incident that slipped by the editors, it went well. Don also had me join them at Camp John Hay in Baguio one weekend to see the sights. Taking this as a personal snub, Timmy was incensed because he wanted to go and take his wife. He was snide and passive-aggressive toward me the entire time Don was on leave, but fortunately, I spent most of that time doing residential security surveys off base. Based on client feedback and my knowledge of the off-base subdivisions, Don felt I was good at conducting the RSS and tasked me with doing the majority.

A couple of weeks after Don returned, my relationship with Timmy imploded during a weekend softball game. I had a summer cold, so I didn't feel well and played worse. Timmy and I nearly came to blows when he cursed at me in the dugout for missing a grounder, causing Major Rich (3rd LES commander) to step in and separate us. I'm not sure why we both weren't dismissed from SPAC and immediately returned to gate duty. While the drama passed, we were never the same toward one another, and I avoided him to the extent possible at work and completely while off duty. I've always disliked hyper-competitive egomaniacs, and my colleague turned exposing his negative attitude into an art form.

Around that time, a new NCO named SSgt Penley Wallow reported from Carswell AFB in Fort Worth, Texas. He had been the Crime Prevention NCOIC there, and reportedly, his claim-to-fame initiative was inventing a metal post with handle hooks on it to prevent dogs in base housing from knocking over garbage cans. I first met Wallow while eating at the Kelly Cafeteria one Saturday morning; he saw my "3rd SPG Sports" shirt and introduced himself, then invited me to come over and sit with him. He laid out his crime prevention ideas for a new program he would oversee, insisting that Colonel Allison promised him the SPAC NCOIC job if he could get assigned to Clark AB.

Hearing all this made me uneasy, especially when Wallow introduced himself as the "new Crime Prevention chief." I didn't know anything about him other than Allison thought highly of him and intended to rotate Don to other duties once Wallow learned the ropes. Criminal threats in the Philippines were vastly different from those at stateside bases. Although Quesnell and Wallow graduated from the same crime prevention courses, Wallow's lack of understanding of the Philippines' special challenges created immediate friction with Don. For starters, he wanted to dismantle Don's programs and get rid of our team. The latter didn't impress most 3rd SPG staff and senior staff NCOs who had seen measurable progress with Don as the SPAC NCOIC.

Colonel Allison listened to the concerns of both NCOs and decided to assign Wallow as a Resource Protection NCO — "an equal peer under the same roof" — and assigned Timmy Tinker to him. While I didn't work "for" Wallow, I was to work alongside him without caveat, which I found difficult, knowing he didn't believe in our programs. Worse for me, Tinker immediately seemed to feed off Wallow's negative outlook toward our current initiatives and started arguing against Don and me. This acidic relationship was untenable and would inevitably come to a head.

The USAF Security Police had a tactical competition called the Peacekeeper Challenge. One had to be in good physical shape and an expert marksman to make the team. The 3rd SPG built an exact competition model of the obstacle course near the base armory; while negotiating the course was not very difficult, this event weeded out many team aspirants. I had trained on our course in my off-duty time and thought it was fairly easy. Already meeting the physical and shooting prerequisites, SSgt Quesnell said I could try out.

I showed up at dawn the next day ready to train. A Philippine-American named LT Almacen was the team captain, and MSgt Funk was the team NCOIC. As expected, I breezed through the obstacle course and made the team. We had a month to work out and refine our skills before traveling to Kirtland AFB, New Mexico, for the competition. I returned to work and informed Don, who congratulated me. He said he would approve my taking leave to visit home while stateside.

The next morning at first light I showed up for training only for LT Almacen to inform me that one of my black friends, Clarence Mangum, would take my place to "racially balance the team." MSgt Funk glanced at me and then looked away, so I sensed he was overruled. This made me madder than hell since I made the cut ahead of the other airmen; I was a better shot, could run faster, and posted a better obstacle course time than every other new guy on the team. Conversely, I really liked Clarence and was happy for him, so swallowed my pride. Told I was now an alternate with almost zero chance of going to the Kirtland AFB competition, I declined further involvement and returned to work.

Annoyed over the reasoning for my relegation to team alternate, Don's protest to our commander went unanswered. Several team members asked me to return in case someone else got hurt and couldn't go. I refused out of principle, plus there were a half-dozen other alternates from which to choose. I don't recall how our team fared at Kirtland AFB, other than it didn't place in the top three teams, except for our K9 team, which I seem to recall placed second. Given the opportunity to turn back time and do over, I would have stayed on the team and worked harder for the experience I could have gained from the training. Lesson learned.

There was a skinny bonehead named Jerry who lived down the hall from us in the barracks. He spent most of his off-duty time getting drunk and picking fights — probably his primary leisure activity since the 7th grade. He had a thick Yankee accent, and I slapped the hell out of him once for saying, "Southern people are stupid, inbred, and racist." His penchant to run his mouth at guys nearly twice his size got him beaten up a lot. He was obviously many fries short of a Happy

Meal, so most people just avoided him. He had an unwise habit of getting bombed and watching TV in the open day room, so we frequently found him passed out on the floor there. Guys liked to take Instamatic photos of him passed out and duct tape them inside the urinals for targets.

One night, Dean Bidwell woke me up to say, "Jerry's passed out in the dayroom. I'm taking your shaving cream too." I sat up in my bunk and asked why. "I'm going to shaving cream him." I laid back down and then thought, Yeah, good idea. I can't stand that jackass. I hopped into a pair of gym shorts and running shoes, grabbed another can of shaving cream from my wall locker, and joined Dean, heading down the hall. We sneaked onto the porch to flank the day room from the outside. Sure enough, Jerry was out cold in a big easy chair with his head back and mouth wide open. Dean excitedly whispered, "Wait — let me go get my video camera," and off he went.

Our prey never stirred for the few minutes he was gone. Now with a video camera in hand, Dean got directly behind Jerry and started filming. I tiptoed in and started covering his face with shaving cream, careful to avoid spraying the expanding white foam into his nose or mouth. Dean and I both shook so hard from laughing silently that we stopped for a couple of minutes to calm down. Finally catching our breath, he prodded me to resume the mission. I again started spraying and soon went through two cans. I stepped back to watch as

Dean circled the guy, filming up close. I was astonished that Jerry could sleep through this and shook all over, trying to suppress the laughter that was about to explode out.

Dean nodded to me to use the third can of shaving cream. I shook my head "no," since Jerry's entire face was covered, and the only place left to spray was into his mouth. I was afraid he'd choke if I did that. Dean insisted and pointed into the guy's mouth. He whispered, "He'll wake up when you do it and spit it out. I gotta get this shit on tape!" Figuring he was right, I shrugged and moved forward with my last can in hand.

I was again shaking violently from trying not to laugh out loud. Holding the shaving cream can directly above our victim's face, I started spraying it in a circular pattern around his mouth and nose while still leaving an air hole. The shaving cream was at least seven inches high on his face and resembled a crawfish burrow one might find on a creek bank. I still couldn't believe Jerry could sleep through us doing this, especially with our snickering that was by then clearly audible.

Suddenly, Jerry twitched a bit, and the sides of the piled-up shaving cream collapsed into his mouth. For a second, the white pile was still, and then it caved in like a sinkhole. Thinking he might choke, I jumped forward to wipe it away when the piled-up shaving cream exploded upward several feet like a volcanic eruption. Incredibly,

Jerry still didn't move! The volcano face erupted again, its mouth and nose now blowing shaving cream all over the sprawled body connected to it. I looked up at Dean, now laughing aloud but still taping.

Jerry sat up then, his head resembling — except for his mouth — a faceless, snow-white mannequin as his hands rose up. He swatted at the shaving cream, splattering it all over his chest and lap. The mouth began roaring very creative curse word combinations as two eyes appeared. I had stepped backward onto the porch, so he didn't see me at all. He looked right at Dean, taping this glorious epic of barracks dickheadery, and lunged toward him. Jerry hit the tile floor like a 120-pound sack of rice and passed out again. I rushed forward and checked for a pulse to ensure his airway was clear. Jerry was fine, just drunk as hell.

At that moment, we discovered someone else must have tied Jerry's shoestrings together prior to our prank. Dean kept taping and declared, "This is why assholes like you should never drink alone." We watched that tape over and over again for months, laughing like madmen at each viewing. The weirdest part is that Jerry had no memory of the event. Because if he had, he was definitely enough of a hothead to have picked a fight over it.

Among the odd aspects of serving in the Philippines was seeing movies that were 6–9 months behind those currently shown in US theaters. Less popular movies were usually released 6 months behind, and blockbusters followed a few weeks later. The justification we heard was that it allowed time to add voiceover translations and/or subtitles before releasing these movies into the Philippine theaters. Whereas I cannot attest to this as fact,

we accepted this reasoning and were glad when the newer movies were finally shown in the Kelly and Bobbitt theaters on base. The Kelly Theater was older and close to Barracks Row, whereas Bobbitt was much newer and next to the BX and Commissary. Both served beer in popcorn buckets to those of legal age, so the Wagner High School kids sometimes pestered airmen to buy them beer. Incidentally, that $2.00 "bucket of suds" cost the same as a movie admission, meaning you could still buy beer and see the movie for $4.00. A large popcorn was also $2.00. While cheap by today's prices, most junior enlisted still complained since their monthly salaries were under $1,000 per month. Naturally, everyone in the theaters (unless pregnant or elderly) stood for the National Anthem that played prior to every movie. No one ever refused to stand, since that could prompt the offender getting tossed out and possibly beaten up. Yelling obscene wisecracks also got guys thrown out, which I witnessed several times. One time, after an obscene sexual remark, another airman found the offender, and a fight ensued. I just sat back with my popcorn and beer, enjoying the movie and the brawl.

I served on the 13th Air Force Honor Guard for most of my tour in the Philippines. Although an additional duty, our services were respected, the duty was enjoyable, and it looked great on our evaluations. The uniform consisted of Class B attire (trousers and short-sleeve shirts), ascots, white gloves, white pistol belts, and chrome helmets. I participated in such ceremonies while in high school, Junior ROTC, so I required no training outside of event rehearsals. Funerals were frequent at the base cemetery for area retirees and Filipinos who had served in US or Philippine cavalry units during the colonial era and WWII. We worked color guards for all sorts of functions on and off base. If leaving the base, we carried wooden drill rifles, and local police escorted our group.

198111: The 13th Air Force honor guard performed during the monthly "retirement and retreat ceremony" at the parade ground. This was taken in Nov 1981. I'm the middle individual in the first file on left side. Half the honor guard members were police from the 3rd SPG. The sharpest one was the airman at far right who was a Junior ROTC drill team stud in high school.

Every month, 13th Air Force HQ hosted a retirement and retreat ceremony at the west end of the parade ground. The latter was nearly 1,000 yards long and 200 yards wide due to its original function as a cavalry drill ground and polo field. We also marched with our flags in numerous parades all over Luzon, including the Lingayen Gulf invasion anniversary in La Union up north and during the Bataan Day observances in Manila. Prior to the Bataan

Day parade in April 1982, we rehearsed a marching rifle drill for several days to get it right for the performance in front of the reviewing stand. When we executed it during the parade, President Ferdinand Marcos and his cabinet members rose to give us a standing ovation.

Many of my SP buddies joined the honor guard, including Dean Bidwell, Richard Palmer, Mike Castleman, and several others. Because the SPs already wore berets, someone proposed that the honor guard should adopt white berets. Surprisingly, the base commander approved, and we got them. The only downside was that non-SPs didn't know how to shape their berets or wear them right. Most looked awful—especially on women with longer hair stuffed up into buns—so we SP members constantly had to fix their headgear.

I felt very badly for Rich Palmer during one monthly retreat ceremony when a big seabird "dropped the bomb" on him. We were standing across from each other at "present arms" during the Philippine national anthem when a whitish meteoric streak flashed by the dark trees in the background and splatted against his M16 and midsection. With his very dark complexion, the white poop really stood out on his arms, almost as if someone had slung a cup of melted vanilla ice cream on him. Someone else also saw it, because I heard a snicker nearby. Other than quickly glancing up, Rich never broke his bearing and continued staring straight ahead toward me. I tried to suppress a slight smile, and it helped that he narrowed his eyes at me.

After we marched off to turn in our M16s for dismissal, Palmer vented his aggravation at the bird as everyone gawked and laughed over his soiled uniform. "Can you believe this crap? With a whole parade ground to shit on, that damned bird picked me! I hate these big ugly geese or whatever they are. Don't niggas and crackers down South eat those nasty ass things? Man, there are even fish scales and bones in this thing's shit!" This went on for a while as he jokingly ranted. He rode in my car and continued talking after we arrived back at the LE barracks. Fortunately, honor guard members were entitled to free dry cleaning, so Rich at least was spared having to ask the houseboys to clean that stanky poop.

Another benefit of honor guard service was getting put on separate rations. This meant we forfeited our "all you could eat" meal cards for a separate subsistence allowance. Not to worry, once we paid at the chow hall, we could still get seconds near the end of meal hours. The reasoning was that honor guard members sometimes missed meals due to regularly scheduled ceremonies. In reality, it was more of a reward for the additional duty, and most of us still paid to eat at the chow hall anyway. However, it did allow us to pick where we wanted to eat and a hamburger combo meal at one of the base clubs cost about the same as the chow hall.

I greatly enjoyed the Airman's Club (aka "Coconut Grove") and NCO Club (aka "Top Hat Club") at Clark AB. Both had good, reasonably priced food and theme events almost nightly. Membership cost $3 per month, and on monthly members-only nights, each club fed us free steak dinners. This was "one steak only" initially, which turned into "all you could eat" if attendance was sparse or they cooked too many steaks. The members-only nights also held drawings for all sorts of prizes, ranging from free bottles of booze to stereos to used cars. I won a Betamax video player during a member night and declined selling it for $500 on the spot to someone who wanted it. I later regretted it, since VHS players superseded Beta in a few years. I kept it for several years and eventually gave it to a cousin who was left with a load of Beta tapes after his family's machine bit the dust.

In late summer 1981, the Airman's Club gave away a very cool 1959 Chevy Biscayne that was in very good condition. It was powder blue and had the classic old "Batmobile" tail fins in the back. One of our A Flight LE guys, Mark Waters, won it. He only kept it a couple of weeks because someone (I think a fighter pilot) offered him a great price, and he sold it. Had I not already owned a car, I would have bought it. At our next assignment, I remember Mark's dad saying he wished Mark had shipped the car home to Texas.

One 3rd SPG armorer was a huge airman named Dan Pardue, a 6'11" and 310-pound former college football lineman from Utah. On several occasions, I saw Pardue walk into the Airman's Club ballroom and stick a funnel in his mouth for his buddies to simultaneously pour multiple beers down his throat. Despite his size, Dan was very laid back and did little to draw attention to himself, since anywhere he went, everyone already knew he was present—one couldn't exactly miss him. I liked Dan and found him to be a pleasant and good-humored guy to be around.

The Airman's Club had a main ballroom with a large bar and several side rooms. One was a gourmet restaurant that served the best Chicken Cordon Bleu I've ever had in my life; it also had a variety of great seafood, Asian, and steak entrees. Another room was the Mirror Lounge—the only disco on base—and it had its own bar. To one side, in the back, was the pool room that had video games and a dozen slot machines. Last was a side window to a short-order kitchen that served decent hamburgers and fries, plus similar entrees. Adjacent to the club in the same complex was a Class 6 liquor store that featured inexpensive booze at heavily subsidized prices. I had several Filipino friends whom I occasionally gave wine and liquor as gifts, so this required calculating the probability of a "show and tell" visit by the Merchandise Control Office. Beyond not exceeding your monthly purchase limit, the common-sense to avoid this trap was keeping near-empty bottles of those brands in your wall locker. If the MCO conducted an inspection, your open bottle proved you still had possession.

The NCO Club had "Mongolian Beef Night" every Thursday. You paid at the door, picked what you wanted your dinner, and dropped it off with the cooks while you hit the bar. When your food was done, they summoned you to get it. I mainly went during the dry season to allow eating out under the stars. Nights from November through May were usually clear and cool, especially in December through February when temperatures dropped into the low 60s.

Both clubs' theme event nights were great fun and featured all sorts of giveaways—mostly free food, rounds of beer, or bottles of really shitty cheap champagne. We enlisted swine always joked that the good booze went to the Officers Club. The booze always flowed freely regardless. The game shows pitting units against one another usually turned hilarious as the evenings progressed. Contestants fortified by liquid courage grew brazen and held little back. "Squadron Feud" nights (patterned after Family Feud) were a recurring favorite for many and were definitely mine. Guys often got mad over missed questions and then cursed and yelled at others for making fun of them;

this sometimes led to thrown San Miguel bottles between participants and hecklers in the crowd. The judges ejected the sore losers and brawlers, so someone else from their units or "mercenaries" from other units took their places. Knowing the Squadron Feud dates in advance, 3rd SPG guys would gather up several teams and try (emphasis on "try") to stay sober to win. Our team usually had one mercenary: a Black Texas cowgirl named Emma Palmer. A fellow honor guard member and Personnel Specialist, Emma was smart and kept the rest of us focused — sort of a den mother to hyper-competitive Scouts. The winning team received vouchers for steak dinners and drinks, plus bottles of booze awarded that night. Even the losing teams won booze, t-shirts, or other prizes to ensure no one left feeling offended, much like how modern kids often win participation trophies regardless of their competition placement. Only these were liquored-up young adults with strong vocabularies far from home. While our team usually placed in the top three of whatever event was held, I intensely disliked champagne, so I traded any I won for something else. I was usually OK with just Lone Star Beer.

There were always a few older local women who hung around at the base clubs. Whereas their base ID cards listed their occupations as "Entertainer," everyone knew they were prostitutes or strippers whose best years were behind them. Knowing the young airmen were neither attracted to nor remotely interested in them, they mainly left them alone and focused their attention on TDY personnel and other visitors.

Sometimes these women turned up in prostitution dragnets in the transient barracks, but always seemed to beat the charges and return to the base clubs within a few days. Incidentally, some of those caught in the dragnets were Pinoy males sleeping over with male GIs. To each his own, I guess.

After my first few trips to the downtown bar district, I decided I disliked the overall scene and seldom went down there. Morals aside, I didn't appreciate the bar culture with its blatant prostitution and sex shows; however, like almost everyone else (including female military personnel), I went along with friends now and then. I saw things almost beyond description—disturbing stuff like occasional bestiality, graphic homosexuality (especially women on women), and extensive sexual activities performed by people desperate to make a living. Americans willing to pay for such entertainment perpetuated it, as it was simply a "supply and demand" equation. Moreover, a large percentage of single American men married local women they met in the bars; roughly half of single enlisted men did, as was predicted by briefers during our base in-processing. This sometimes proved a win-win for both parties in these relationships. That aside, I must emphasize that many single Americans married college-educated professionals whom they met on base or elsewhere in the local area. Don Quesnell's wife, Belle, was the executive assistant to a colonel. Airman Castleman married a bank employee. Another guy in my unit met and married the daughter of a farmer who raised sugar cane and hogs; he left active duty to join her family's farming operation. There were many cases like this, especially among the NCOs and officers. Late in my tour, my barracks roommate Dean married a colonel's maid whom he met during a walking patrol in base housing. Their blossoming romance nearly got him in serious trouble.

To digress slightly, the story of Dean meeting his future wife in the senior officers' housing area (the Hill around Mactan Circle) was quite amusing. I was that area's patrol leader the day they met, and the desk sergeant couldn't reach him by radio. After a period of searching, I found him walking his beat; he attributed the lost contact to a drained battery, so I let it go at that. Afterwards, he constantly volunteered for back gates in hopes of getting a walking patrol in the Hill area the following day. A few weeks later, the colonel's wife returned home for lunch one day to find Dean at their dinner table having lunch with their maid. Colonel Allison was one of their neighbors, so word traveled fast. While Dean got in no serious trouble over it, that was his last walking patrol in the senior officers' housing area.

Anyway, I felt sorry for the young Filipino women working the bars. According to those who understood the business, most were either pressured or compelled to enter it for lack of a better way to make a decent living. This was a constant challenge in a country where unemployment was as high as 30% at the time. Women returning home to their provinces often had nice clothes and spending money; they told other young women about the great jobs they had in Angeles City or other cities and how they met "rich" foreigners. Faced with hard lives working the rice fields or doing manual labor at home, it often took little to persuade good-looking young women to head to the cities. While some of the promises were true, leaving out the darker aspects was a shock to many girls once they discovered the expectations. The vast majority lived upstairs in the bars where they worked, and few had the means to return home even if they wanted to. From all accounts, many were assigned bartending duties until ready for more lucrative (or lurid) employment duties. Incidentally, this occurs in a lot of US and European cities as well.

Officially, prostitution did not exist. If a man wanted to take a woman from a bar, he had to pay the bar a fine ("bar fine"), and beyond that, whatever occurred was supposedly between them. Stories of beautiful women charging significant sums to visiting sailors at Olongapo outside Subic Naval Station were well known. In port only a short time with money to burn, few sailors had a concern about paying the inflated bar fine rates. The same applied to visiting airmen at Clark Air Base. This obviously wouldn't fly with permanent party airmen and sailors, so their going bar fine was usually around 100 pesos (roughly $10 USD). For men and bar girls desiring to remain together exclusively, bar owners had a novel arrangement known as a "steady bar fine" that cost the man a single charge of 1000-1200 pesos.

I knew many Americans whose relationships with their wives began by paying steady bar fines. While many people back home were openly shocked when I later described the concept, I personally didn't have an issue with it. To me, this was simple: If the two parties involved were OK with the arrangement, that was their business. To those who complained about it, my standard answer was "Then don't participate in it." The Philippines was a Laissez Faire Land where we were expected to leave everyone else alone to freely pursue their own paths in life.

That aside, many bar girls got a good deal by marrying Americans, so it was viewed as somewhat culturally acceptable. Half of the US servicemen assigned to the Philippines married Filipinas, and my group was no different. Of the ten guys in my SP Academy class assigned to the Philippines, eight were single, and four met their future wives in downtown bars. Whether they were saints or hookers wasn't my business, and I didn't care enough to ask. Decades later, the majority with whom I've stayed in touch are still happily married to their first Pinay wives and busy being grandparents. Ultimately, I just didn't take to the downtown bar scene and mostly spent my off-duty time at the base clubs, played sports, or road tripped all over the island of Luzon. I often wish I had spent less time playing sports and more time seeing the country.

Interestingly, most of the downtown bars had softball teams that played in a league in Astro Park right outside the Main Gate. Most bars' team rosters consisted of guys who dated women from those bars or whose wives were former employees. I never joined any bar teams, yet did fill in as a substitute player for a couple when needed. While the teams were competitive, they had a lot of built-in fun and performance prizes too. Whereas winning a game entitled all players to a round of beer, hitting a double or triple was worth a beer; the same applied to double plays on defense. Since hitting a home run might win a batter a free bar fine, many guys predictably struck out or popped up because they liked swinging for the fence. Drinking during games also contributed to a disproportionate number of outs.

One Saturday, I was subbing on a friend's team and hit two home runs. It was a clutch game, so the bar owner issued me two free bar fines and told me I could redeem them any time. I sort of had a girlfriend at the time, so that wasn't happening. The younger sister of another friend's wife worked at that bar, so I asked if I could donate the two nights to her so she could go visit her family up in La Union province. I sensed my gesture genuinely surprised the bar owner, and he agreed. Over four decades later, it bothers me that I can't remember any of their names. However, I do remember that young woman's tears of joy when told she could go visit her parents and grandparents the following weekend. Several of us passed the hat to buy her a round-trip bus ticket. It was a win-win for everyone involved and perhaps more so for my own sense of direction in life. Like the Grinch, I was starting to grow a heart — maybe a little.

Many colleagues pestered me over my reluctance to seek local female companionship among the Pinay girls. Some even suggested I might be gay, definitely not the case. I had viable reasons, and the most compelling was that I had no desire to settle down before I completed college. I was dead set on serving my four years, leaving active duty, and joining a reserve unit while earning my bachelor's degree; I also wanted the option to enroll in ROTC and earn a commission. Marriage and supporting dependents would complicate and possibly eliminate those options. Remaining single afforded me sufficient flexibility to pursue multiple career and life options.

Another issue for me was psychological: The bar girls downtown were often aggressive, and the older ones were both experienced and intimidating. Some were masters at negotiating situations in their favor, especially when the men they pursued were often lonely, drunk, or both. I observed many young and relatively inexperienced Americans getting cooed, cajoled, and manipulated into ill-advised relationships with no realistic way out other than to break contact (best case) or hurt the women with whom they were entangled. In my mind, avoiding these situations altogether ensured I stayed on track to achieve my educational and career goals. Did I blame the girls for going after these American men? Not at all, for this was the pathway to a potentially better life for them and their families. I emphasize "potentially" because in some cases, the girls married guys who were already low performers with little ambition and would remain so; indeed, men who are lazy at work tend to be lazy at home too. Over the years, I've met plenty of foreign-born women who ditched such men and remarried other Americans. The pendulum always swings both ways in this regard. And by the way, the preceding observations apply as well to Americans of both genders. As an older gentleman advised me decades ago, "Boy, if you want to date and mate well, be a well-shined wingtip — not a dusty loafer."

I will share another observation. Although closeted to remain in the US military, the gay and lesbian service members had it made in places like the Philippines. Being gay or lesbian wasn't a cultural taboo there, and financing accommodations off base to engage in those lifestyles was relatively inexpensive, especially when multiple people chipped in to lease a place. I honestly didn't know a single person over there who was discharged for homosexuality. I contend this was due to unit commanders' far greater concern with diligent duty performance than who or what their troops were shagging.

I'm obviously omitting the lurid stuff I witnessed, things that I "saw once for the experience." Just to name one notorious place, the sexual debauchery scenes I saw at the "Nipa Hut — Human Zoo" were the stuff of legend. When these shows ended, attendees could go onstage to "congratulate and tip" the entertainers before exiting through the Space Ship hatch that led to an exit back to the street. Over the years, whenever I shared stories about the wild times and hedonism I witnessed just beyond the gates, almost no one ever believed me unless they were also there. To quote an older friend many years later, "Sometimes it's best to keep these experiences under lock and key in your brain. Most people neither understand nor can relate to such times, much less appreciate the situations and cultures we lived through. Don't try to impress them with what you know. Just smile and keep these things to yourself until you're ready to write professionally about them." Even now, there are things I contend that are best left to one's imagination.

Aside from the regular Cope Thunder exercises, we participated in numerous other base-wide alerts and exercises. Most followed a set script: Recall security personnel, rope off flightline areas, set up security cordons within aircraft parking ramps and around the flightline, and start generating aircraft to pound targets at bombing and gunnery ranges or dogfight over the South China Sea and Pacific Ocean. Everyone worked 12-hour shifts, and I always seemed to draw the day shift where I was baked by the sun or soaked by cold rains. Sometimes I was pulled for QRF duty with the EST teams, mostly during the latter 12 months of my tour.

One exercise I was assigned to the M60 machine gun post (part of the DSS screen) at the south end of the runway; however, my M60 was "simulated" due to the rain, so I got posted alone with only my M16 rifle. When I opened the concrete pillbox door, a very large brownish-gray snake greeted me, so I slammed the door and let it have the place. Beyond that, the bunker had several inches of standing water, so I just sat on its concrete roof.

Lone Star was hosting the Chief, Security Police for PACAF during that exercise and stopped by for a post check. When he asked why I was on top of the bunker, I told him to check out the snake inside. He opened the door and remarked, "Man, that's a damned big snake! I'm surprised you didn't shoot it." I replied that had it attacked me I would have. He called Animal Control on the radio and told them to come get it. When they asked what type he thought it was, he replied, "Well, it's big enough, I don't want to find out." He handed me my boxed lunch, and they left to go see someone else down the line.

Animal Control soon arrived and caught the snake. It was a nonvenomous python about 8 feet long. The Pinoy who caught it laughed and said, "No big deal — just a juvenile." He added, "Mature ones grow to 20 feet or more." I thought, Yikes, I'm never going camping here! Another time I was on the same post, and so many F-4s took off directly over my head that I thought I might go deaf. If an aircraft had an emergency and ran off the end of the runway, the bunker was almost in the direct path. Fortunately, none ever did. As it stood, they were close enough passing overhead that I could feel the hot breeze and taste the jet wash fumes after they flew past. Between the blazing sun and pouring rain, and the noise, that DSS post was always a miserable 12-hour shift.

Captain Hurt from another LE flight brought out my lunch one day while I was on a DSS post and talked to me about college ROTC and earning an SP officer commission. He was a solid leader, and many times I wished he was the A Flight LE shift commander. I knew at some point I would leave SPAC and return to regular LE duty. I also liked 1LT Trombley, a much older mustang officer who was an NCO for many years. His wife was from Hawaii, so he liked to take remote Korea assignments to obtain "base of preference," then get assigned back to Hawaii or Clark AB. Years later, I was told he served most of his career in PACAF.

The summer softball season came and went, followed by track and flag football seasons. While a starter on third base, our 3rd SPG softball team was average. We won over half our games, then were eliminated early in the base tournament. However, our base track team was fairly good and competed mostly against itself and local teams. I was assigned to the relay sprints (880-yard, one-mile, and two-mile) and 330-yard hurdles to save our better sprinters for the pure sprint events. Clarence "Jei" Richardson, who ran track at Troy State, was our team captain and mentored me. I was our only white sprinter; the rest were black NCOs who loved referring to me as their "token white boy." I got along well with them, and they took me to the dance clubs downtown — Third Eye Disco and a few others. Yet I could not escape a stark reality: whereas my 9.9-second 100-yard dash and 55-second 440-yard dash times were competitive for the high school level, I was average against these better-developed older men. With multiple sprinters running the 100-yard dash in the 9.5 - 9.7 seconds range and their 440 dashes in 51-54 seconds, I was not destined for track stardom at Clark Air Base. Plus, I hated training for distance (6+ miles), and our coaches insisted we run distance to build our endurance. After a while, my feet and joints hurt, and I looked forward to the season's end. I completed the one full season I was in-country and then moved on to the next sport.

Next was flag football, and it suited my tastes better than any other sport. The 3rd SPG had a good team, and we ended up in second place out of around 30 teams. Like most large-unit teams, ours had a lot of former college players; most dropped out due to academic reasons or marriage, while others enlisted after graduation. We had some awesome players, including Willie Massey (tailback and receiver), Dwayne "Tex" Lowery (linebacker), Clarence Richardson (quarterback), SSgt Jones (line), Earl Ford (all-purpose player), Don Adams (lineman), Mike Tipton (lineman who played at Alabama), and a black TSgt lineman nicknamed "Dawg" after his former Georgia Bulldogs team. We also had Dan Pardue, a former college lineman from Utah and the lone man I've ever known who could carry an M2 50-caliber machine gun by himself.

With my speed and high vertical jump, I played outside linebacker and special teams. Willie, Earl, and I were the fastest men on the team, but they played primarily offense and rotated in as defensive backs. Given a choice of where to play, I chose to play defense only as a linebacker. With big linemen in front of me, I could look for openings and blitz both inside and around the corners. Our coaches assigned me several blitz packages, and I loved disrupting opponents' backfields. I broke through gaps and charged quarterbacks and ball carriers, plus took every nasty shot I could get away with.

Don Quesnell attended some games and noted my aggression with growing concern, despite team members and our NCOs, and officers urging me to play even harder. One game, Don walked up to the fence at halftime and talked to me; I was so fired up that I didn't remember a thing he said afterwards. In one game, I sacked the opposing quarterback six times and sent his right guard to the hospital with a concussion. Those violent nighttime games under the lights were an intense adrenaline rush.

However, there was always a price to pay via retribution. In the hard-fought win against Camp O'Donnell, I was hit by a blind side block and suffered through a three-day headache. I lost half a front tooth in another game, requiring dental work. In yet another, I was elbowed in the nose so hard that I couldn't breathe through it; it bled off and on several days, and the doctor diagnosed it as a deviated septum. While I stayed constantly scraped and bruised throughout the season, I missed work only for medical or dental treatment, and my tempo never decreased. On the other hand, our unit leaders attended the games and frequently brought free beer for our post-game parties — a larger motivator than the cheering crowds. We were on a roll until the base championship game against Supply, and lost by only six points. Colonel Allison attended most games and served us Lone Star beer from iced coolers in his command car's trunk. He did this regardless of a game's outcome. After the championship loss, he saw I was oozing blood from a cut mouth, so he handed me a chunk of ice and a Lone Star. "Here, Clark, put the ice on your mouth and drink the beer. I really enjoyed watching y'all play this season." Our commander showed genuine care for us, and we appreciated him for it.

Because flight lessons and aircraft rentals were relatively inexpensive, I took pilot training through the Clark Aero Club. The planes were standard Cessna 172s (model M, I think) and easy to fly. I knew most of the flight material due to three years in Air Force JROTC, and I passed the student pilot exams. One fighter pilot (a major) who knew my father's family offered to give me most of my flight lessons for free, so I jumped at the opportunity.

My first flight lesson was somewhat amusing. The major flew us up to about 5,000 feet and gave me the yoke. After a few minutes, I adjusted well and was keeping it level and steady as we flew over Tarlac Province, northwest of our base. He then told me to pull back and climb, remarking, "See how the rice and sugar cane fields are getting smaller?" I acknowledged and he directed, "Now descend nice and easy to 3,000." A few minutes later, I leveled off and he asked, "Did you notice how the fields got bigger again?" I sort of smirked as I replied that I did. He then ordered, "OK, now take us back up to 5,000 again, slow and steady." I wondered whether this was some sort of test or if he had an actual teaching point for me. Once back at 5,000 feet, he said, "Now climb hard without losing control." I gunned the throttle and climbed. "Nose higher — go steeper." So I did and began to feel the Cessna stalling. "Push it — hard climb! Straight up!" Of course, we stalled and the plane began falling like a rock as I fought the yoke to turn out of the fall. The major yelled, "You see how the rice fields are getting bigger? Don't ever do that again!" He grabbed the yoke and pulled us out of the spiral in a few seconds. We spent the rest of the lesson talking about stalls and repeating the exercise. After learning how to recover from stalls, we were confident that I could do it again with no assistance. Yet he warned me to never stall just for fun, as they were dangerous.

By the time I left Clark AB the next summer, I had my license and figured I could resume flying when I got to my next assignment. Perhaps out of naivety, I didn't consider how expensive plane rental, fuel, and insurance would cost me once back in CONUS — about 10 times per hour what it cost me at the heavily subsidized Clark Aero Club. Realizing I could not afford to fly anymore, I let my pilot's license expire with no regrets. Beyond that, working swings and mid shifts left no time for it anyway. I was still saving for college, and that was my highest priority.

Occasionally, Clark units would send teams or individuals to Camp John Hay (CJH) in Baguio and other installations to conduct training and assistance visits to joint or allied forces. Commanders and senior staff members often used "supervising" them as a pretense to travel separately. Translation: They played golf, went shopping, or lounged on nearby beaches like a bunch of walruses. Other than visiting the installation command team, we rarely saw them. Anytime I was at CJH, I rented golf clubs and played the course. Due to the mountain terrain, this meant teeing off on one hill toward the green on another hill, thus I unpleasantly discovered what a pro golfer once described as playing "more of an obstacle course than a golf course."

During one TDY to CJH, I taught crime prevention techniques to the DoD Police Department. My senior NCO host invited me to join him and his local-born wife for dinner at the main club's dining room.

I had eaten there before, so I knew the food was top-rate and reasonably priced. White table linens and candles made for an excellent ambiance. Other than my host and me, there were almost no Americans in the place, so we stood out. As we ate, I sat with my back to the door across from my host and his wife. Suddenly, their demeanor changed, and his wife's eyes narrowed as she tapped his arm. I knew from their reaction it wasn't good.

He said, "Whatever you do, don't turn around."

Several senior-looking Americans had walked in with beautiful young women who were most definitely not their wives. One was talking loudly before another told him to shut up and quit making an ass of himself. Cutting my eyes toward the bar, I noticed they were ordering drinks before getting seated, and all appeared in a celebratory mood. My host recognized them and seemed extremely uncomfortable, again whispering for me not to turn around. He was concerned the other Americans would spot us and come over to talk.

And that didn't take long. I heard a voice saying hello to my host and looked up to find a tall stranger looming over our table. He made small talk for a few minutes without ever addressing me, and then went on his way. My hosts were obviously upset and told me to keep quiet about what I had seen. As a young enlisted man, it unnerved me to have this kind of knowledge about senior American personnel.

The next day, I rode the bus back to Clark. A couple of days later, one of our senior NCOs summoned me to meet him outside; he told me I was to return to Baguio on a "confidential" mission. Growing hostile when I inquired about the details, he refused to divulge anything other than I was to take a sidearm, get on a waiting chopper, and report back to him immediately when I returned. So I picked up my pistol at the armory, reported to the flight line, and got on a Huey helicopter bound for the Baguio Airport. The pilots didn't talk to me other than to warn, "Buckle in and don't bother us." Obviously, this was a flight training mission for them, yet somehow included me. During the flight, I watched the green rice paddies and coconut palms pass by underneath us. Farmers waved at me and I waved back. A couple of kids flipped us off, which I found quite amusing. The mountains grew larger as the helicopter wap-wap-wap-wap-wapped along, and I wondered what I was supposed to do when I arrived.

We soon landed at the mountain airfield, and a Baguio police jeep picked me up. The cop spoke English and told me I was to pick up a package; he said he would take me there and return me directly to the chopper. We drove to a really nice older hotel in town where the desk clerk gave me an upstairs room number and pointed toward the stairs. A bit uneasy after ascending to the top floor, I drew my pistol as I walked down the hallway before gently tapping on the room door. A young woman opened it and smiled. She handed me a gold ring and giggled, "Bye, cherry boy!" and then slammed the door in my face. My first thought was, Why does everyone assume I'm a virgin? followed by, What the hell just happened here? I scurried to the stairs and holstered my pistol on the way down. The Filipino cop at the door motioned for me to hurry back out to the jeep. He returned me to the airfield, said goodbye, and saluted me. Puzzled by the salute, I rendered it and climbed aboard the bird. Moments later, the Huey was airborne and zooming past the big cliff at the end of the runway.

All the way back, my mind raced and my stomach rolled. The wedding ring obviously belonged to an American — likely one of the men I saw at CJH — and now I was involved by proxy. While I didn't know their identities, it was a safe bet they knew mine. Upon landing I reported to the senior NCO who sent me and handed him the ring. He warned, "If you ever breathe a word about this, you could go home in a box. Do you understand me?" I nodded. "OK, turn in your pistol and back to work you go." Almost as mad as I was scared, I turned in my pistol at the base armory and said nothing to anyone there or when I arrived back at the SPAC office.

A few weeks later, I heard a rumor that a senior officer on TDY to Clark had forgotten his wedding ring in a Baguio hotel, and one of our SPs was sent to retrieve it. I went to see the senior NCO who sent me on the mission and assured him I never said a word about it. He said, "I know you haven't, and that's why you're safe. You're a real shithead sometimes, but a competent one, and that's why we trust you." He emphasized the word "we." So now, although clearly protected by God only knew who I had to watch my back and not trust anyone.

Toward the end of the monsoon, we had a 3rd SPG air base defense exercise at Camp O'Donnell near Capas. This was the site of a notorious POW camp during World War II and was now home to the unit managing the Crow Valley Bombing Range. We bused in, drew a load of blank ammo and pyrotechnics, and humped out to the perimeter. We set up a perimeter defense and dug in to repel an attack. After defending against a squad-sized attack late that afternoon, a bunch of us were pulled for a counterattack. Our counterattack stopped at the perimeter, and I was left exposed in the open with only a ditch for cover. An umpire ruled most of us "dead," so we returned to the defense line and ate C-rations.

When an armored vehicle crept past us on the outer road, an SPS guy handed me his M79 to fire several orange chalk target rounds at it, as did several other grenadiers down the line. By the time passed in the opposite direction, its sides were covered with orange dust as if heading to a Tennessee or Auburn game for a tailgate party. Afterwards, we sat around on our tails until nightfall and then moved out again to set up forward positions nearer the perimeter in the darkness. I had no idea who else was in my squad other than the guys next to me, since a senior NCO just grabbed up whoever was available and told us to follow.

Having no briefing on the operational concept beyond the mission to "hold the line and wait for orders," we sat in the meter-tall grass and listened. A Michigander named Hans Lang and I were posted together until a senior NCO replaced him with SSgt Quesnell. Don soon moved to a different point ten meters to my right, overlooking a deep washout, and disappeared in the grass. Around midnight, I got hungry and used my P-38 to open a ration can. Due to the no lights restriction, I had no idea what it was until I took the first bite: "Ham and Eggs, Chopped." Of all the meals I could have opened, I somehow got my favorite. Due to its heinous looks and smell, this meal was called "the fried brains of unwanted pets." While good when heated up and doused with hot sauce, it was still OK cold. Either way, it was better than going hungry. Don came back after a while and said I could take a quick nap if I was tired. I did and awoke sometime later to the sound of blanks being fired. I hopped up and started firing toward the perimeter. Minutes later, a senior NCO, along with SSgt Jim Hill, appeared from behind me with a squad of SPS guys and asked what I was shooting at. I said I was returning fire at whoever was shooting at us from the perimeter. They took off in that direction, and I didn't see them again. Don and several others suddenly appeared, and he said, "Time to go. We're relieved." I followed them back to the main defense line.

We heard firing around the perimeter for the next few hours. We mostly sat around with an M60 team and watched the occasional flare light up the darkness in the distance. The only thing breaking the monotony was participating in an ad hoc patrol around the defense line. As a faint light grew in the east, all patrols and listening posts were pulled in for the "Mad Minute." We received the order to open fire and expend all our ammo to wipe out any bad guys who infiltrated the areas forward of our line. Dozens of flares went up along the line, and a huge swath of terrain was bathed in an intense flickering light for several minutes.

At daylight, our patrols resumed. Told to police the area of trash, we picked up all the flare parachutes and other debris we could find. Told I could keep the parachutes, I stuffed them in my rucksack to take back for Dick (the tailor), and he seemed tickled to get them. I also picked up several dozen unopened C-ration cans that guys had abandoned around the perimeter and took them back to the barracks in my rucksack. Little did I know, but someone had swiped ten or so cases of C-rations during the night. I was later questioned about it, but the rations I found were individual cans, which those I patrolled with confirmed. We got bused back to the base around noon and spent the rest of the day cleaning weapons.

Among the stray C-rations I picked up was a can of pound cake stamped "Sep 1945." The DoD warehouses inspect pallets and cases of rations occasionally, so obviously, the case this one came from was deemed safe to eat. Now, 36 years later, I wondered if it was still edible; it was among the items I brought home to my father when I rotated home. He had served during 1950-53, so he could have eaten pound cakes from the same manufacturer and case lot numbers. We opened the can when I returned to Tennessee in July 1982, and the cake was indeed still edible.

When General David Jones, Chairman of the Joint Chiefs of Staff, visited Clark AB in late 1981, our and the Philippine Air Force's honor guards met him at the plane. We lined both sides of a long red carpet and were to present arms with our unloaded M16s as he passed. For some reason, there was so much noise on the flight line that only the guard members near the plane heard the command to present arms. Instead of a single movement, our arms came up in a ripple effect down the entire line, a spectacular and serendipitous outcome. A command car whisked the general away to Base Operations for a short visit before he toured the rest of the base. We SPs on the honor guard now loaded our M16s to serve as his security escort during his base tour. It was uneventful until we neared the transient barracks by Lily Hill.

I was riding shotgun in the lead jeep and recognized a K9 sergeant in civilian clothes standing on the lawn with a samurai sword held up in a salute position. We pulled over to investigate what was happening. Glancing up at the barracks, a large sign — clearly painted on sewn-together bed linens — rolled down from the top floor's outdoor railing. It read "Welcome GEN Jones, CJCS" and under it "Clark County Sheriffs." As it unfurled downward, a line of men dropped their pants and gave the general a 21-moon salute. I turned to see the general laughing as his vehicle passed, followed by the next vehicle containing a visibly upset base commander. The third vehicle was a staff car with our 3rd SPG commander turning red from either laughter or anger. We later learned Colonel Allison was on record as being "highly irritated and embarrassed by this shameful incident," yet off the record, he thought it was very funny. I later heard that although the wing and base commanders nearly had an aneurysm over the incident, the visiting general found it amusing. He allegedly told the offended commanders he had never before received a 21-moon salute, asking that they not discipline the culprits. As expected, no one saw anything or knew who the perpetrators were, and every K9 guy swore they were set up by people in the adjacent Hospital or US Navy barracks. While anyone with five brain cells knew the K9 troops were "behind" it, only the NCO with the sword was implicated, and he reportedly got off with a wrist slap. Many years later, as Colonel Allison and I reminisced about the 21-moon salute incident, he said, "You know something, I always regretted not getting one of those at my retirement ceremony."

As a preface, one can always tell in which ocean a big storm occurs by its name. Whereas Atlantic and Indian Ocean storms are respectively called hurricanes and cyclones, in the Pacific Ocean, they are called typhoons. In November, the Category 4 Super-Typhoon Irma slammed into the Philippines and made landfall against the mountains on the eastern coast of Luzon. All nonessential personnel were given a day off prior to its arrival to prepare for the typhoon.

Barracks occupants with private vehicles discovered why there were big steel eyebolts in front of parking spaces, as the barracks NCO offered us chains to secure our cars' front axles. Those with motorcycles and bikes were allowed to bring them into the barracks' common areas. Barracks houseboys cranked shut all the aluminum window louvers and slide-bolted most exterior doors shut. We could still come and go, but most elected to hunker down and enjoy the AC in our rooms until we lost power.

I had just moved downstairs, so the heavy rain hitting the barracks and trees was akin to white noise. Although I slept through most of the storm, I kept hearing big thuds, and it took investigating to determine the sound was from coconuts flying off the trees and hitting the walls of our barracks. As the storm passed, the winds shifted, and after a long day of being cooped up, I was grateful when it finally passed. The typhoon did minimal damage to the base infrastructure, and we never lost power other than the normal rolling blackouts (1600-1800). However, it made a huge mess (fallen limbs and leaves) for the houseboys to clean up. The typhoon did a great deal of damage in the provinces farther east and to the north. The large V- and

U-shaped drainage ditches on the base filled up and overflowed in many places as the floodwaters surged off the installation.

Late in the day, after chow, several of us drove out to the Abacan River bridge outside Friendship Gate to see the spectacle of the flooded river. To our horror, a human body suddenly appeared and rapidly floated past us downstream, obviously a person swept away by the river somewhere upstream past Sapangbato. I felt completely helpless and useless watching that dead body shrink and finally disappear into the distance. Several hundred people perished during Typhoon Irma, so this one was just one of a reported 200 or so other victims.

After flag football ended in late 1981, I started to lose my grip on things, and it was a long time before the next sports season started. My lone enjoyable non-duty activity was working honor guard details. A lot of guys were returning to CONUS on leave, and I wasn't seeing anyone, so I grew lonely. Over Thanksgiving, I grew depressed and increasingly homesick. Now, in the lull between team sports seasons, I started partying late (even during the work week) and getting into brawls. At one point, I slept through my alarms for three mornings straight. With the dry season at hand and the elephant grass drying up, I was discreetly invited to join several K9 handlers on a legendary annual pursuit — burning off the dry grass and brush. My friends knew I was struggling, so they suggested I "do something" to get me in better spirits. How this transpired was purely accidental.

Earlier that year, some K9 patrolmen caught me cutting down a clump of banana trees with an Ayta-crafted samurai sword next to a barracks known for bicycle theft incidents. A witness to one theft said he saw a local emerge from the Navy barracks, snatch an unlocked bike, and then ride off eastward. Once thieves reached the protective concealment of the tall elephant grass, bamboo, and wild banana copses covering the large expanses of the base beyond the cantonment area, they were gone. So I embarked on a one-man crusade to eliminate the bad guys' concealment wherever possible. The K9 patrolmen evidently sensed I had a point and aided me in removing low-hanging limbs and banana clumps around other barracks near theirs, cutting "fields of fire and pursuit lanes for our dogs," they reasoned. On one occasion, a couple Negritos assisted us in cutting the larger low limbs, and I even hauled a car trunk load of 3-5" thick branches out to Negrito Gate for them. It was free wood, and they would definitely find a good use for it.

Our K9 colleagues were constantly harassed by intruders infiltrating from our perimeters. Clark Field had concrete walls along parts of the perimeter bordering the smaller towns and base housing areas. Beyond those, it was a wire fence in name only, where the fence was not kept under constant observation; thieves cut down the wire nearly as fast as our engineers could erect it. In some places, nothing but concrete posts stretched for several hundred meters. From patrol reports (i.e., encounters and small hut discoveries) of intruders' presence, we knew crooks hid out in the washout areas and even lived there during the monsoon. Chasing potentially armed aggressors into the bush was dangerous; SrA Robert Gray was killed pursuing some in the late 1970s. As such, the K9 and horse patrol teams longed for the monsoon rains to cease so they could burn off the grass. The annual monsoon usually ended around Halloween, and within a month, the dry brush and elephant grass were ready to burn. While our SPs were disciplined about when to begin burning, they held nothing back once the torching began — a pyromaniac's dream come true.

My K9 friends trusted me enough to approach me about assisting them with the annual fire-fest. They insisted that if the job was done right, we would deprive the criminals of their hideouts and make the entire base safer until the monsoon resumed the following summer. The fires typically burned until they reached the eastern perimeter road; even if the fire jumped the road, it stopped at the adjacent MacArthur Highway and railroad bed. The base fire department easily controlled any fires that might backtrack to the mown grass along the runways. I agreed to join them, an offer I couldn't refuse. With illumination (slap) flares readily available at the Nepo Market for a nominal fee, I discreetly bought a few and hid them. I also bought several dozen books of matches off base.

Two weeks later, after evening chow, a K9 friend saw me in the parking lot. He said to meet him behind the transient barracks at 0015 that night and parted with, "Wear jungle fatigues and bring a camo stick for your face and hands. Bring a full canteen. If you have a pistol, bring it too." My first thought was, How does he think I might have a pistol? My second thought was, No thanks, my pistol isn't coming. That night, a jeep arrived at the pickup point, and we headed toward Mabalacat Gate. Two NCOs in jungles and face paint were with him, both armed with pistols; evidently, I wasn't the only SP who had shopped at the Nepo Market. About halfway out to the gate, my friend said, "You go with them and do exactly what they say. Hop out when I slow down. Be careful and have fun." Assuming we would walk back to the armory area, I didn't bother asking about the exfiltration plan. A few minutes later, we were out of the jeep and on our own in the dark.

We hiked off the road a couple of hundred meters for a listening halt. I barely knew the two NCOs, but was aware one had served in Vietnam. During our halt, I looked up at the stars and determined we had traveled almost due east, so we were heading to the washouts on the north end of the runway. It also dawned on me that while I brought the flares, I had forgotten my canteen and was going to get thirsty. Moving again, we eased through the very tall grass along what appeared to be a trail for what I estimated as half a mile. Eventually, one NCO handed me his satchel of flares to hold and told me to stay put until they returned. I was surprised they just left me, but they seemed to know their way around. I sat there in the dark like a stray dog dropped off in the middle of nowhere, listening for footsteps and watching the stars. In 20 minutes, one returned, followed by the other a few minutes later. Although both had fired several flares, I was upwind and never saw any lights due to the flightline's ambient light to the south. They liberated the other flares and told me to follow.

Off we went again and soon arrived at the edge of a large washout. As expected, they left me alone again. I figured out they were firing the flares down into the washouts for some sort of light discipline. They soon returned, and we resumed our foray in a different direction. We repeated this procedure several times until we were out of flares. By now, I could see light glowing brightly through the elephant grass in several directions and hoped we were done. After another hour of skulking along and setting additional fires to develop (lit cigarettes folded into books of matches), we headed west toward the horse stables. We met two mounted patrolmen near the Mabalacat Road, and they said we were clear to use the stables for cleanup.

Once at the stables, we cleaned off our camo paint and awaited pickup. Soon, a K9 jeep arrived and dropped me off behind the transient barracks. When we got there, one of the NCOs warned in a menacing tone, "What happened out there, stays out there. Do you understand me?" I said I did. "Get back to your barracks and don't even think about going back out there to the washouts." As I drove back to my barracks, I saw a slight glow in the sky beyond the flightline. Once back at 7504, I climbed the ladder to the roof and was thrilled to see that the expanse of grass in the distance was an expanding sea of fire. Although almost 0500, by that time I was so excited I couldn't sleep and ended up driving to the chow hall at 0630 for breakfast. I was happy the bad guys had now lost their favorite hideouts until the next monsoon season arrived.

My friends were right. Those fires burned for several days, and many more fires popped up in other brushy areas. I correctly guessed under the smoke cover that other K9 and horse patrol troops joined the fire-fest to get it over with quickly. In the ensuing days, other areas around the perimeter mysteriously caught fire. Local farmers routinely set some fires to burn brush and weeds, so a few probably got beyond their control. The assessment that the base engineers and fire departments would let the fires burn out on their own was correct, so the K9 guys knew what they were talking about. Within a week, most areas around the perimeter and north runway approaches were blackened, and the intruders' concealing underbrush was gone for the season. One morning later that week, I remarked to SSgt Quesnell that I thought it was a great idea, quickly adding "whoever did it." He looked up at me for a moment as if studying my eyes, and then said he guessed so. Although he didn't say it, his accusing look told me he already knew. Dan the Pyro was in the house!

On Halloween, I was tasked with handing out candy to trick-or-treaters in the Lily Hill NCO housing area. It was sort of a fun duty, and I was joined by other SPA members and numerous off-duty LE guys from other shifts, most of whom were distributed to other housing areas. The kids mobbed us for candy, and I must have handed out 50 pounds of the stuff, much to the chagrin of the base dentists. Of course, the Public Affairs team showed up and took lots of photos, including some of me with the Filipino kids.

When a photo of me handing out candy appeared in the Philippine Flyer newspaper, many LE peers noticed I wasn't wearing a weapon. Naturally, a few dickheads collected copies and posted the clipped-out photos with snide captions on latrine doors and above the urinals in the barracks. "Airman Clark f—king off as usual." "Does Crime Prevention Clark make you feel safer?" "Oh, Danny boy, where oh where's your weapon — up your ass?" "Wunderkind molests local school kids." While I now wish I had collected these flyers to laugh at in later years, they annoyed me at the time, and I destroyed every one I could find.

Halloween was always a big costume event at both the Airmen and NCO Clubs. Of course, I waited until the last minute and didn't have a costume. Several friends tried to get me to borrow a dress and go in drag, but I wouldn't have gotten caught dead like that. I decided to improvise with bed linens and go as a ghost. I just cut a hole in a bed sheet to stick my head through and cut two holes in a pillowcase — voila!

We went to the club and I didn't think much of the stares at first. I went to the bar for a beer. Someone tapped me on the shoulder and I turned around to see Palmer, my black friend from Chicago, standing there. "Rich, how's it hanging, man?"

"Have you looked in the mirror?" he demanded.

"No, why?"

"You look like the damned KKK and I'm the only thing standing between you and a serious ass kicking by a whole bunch of people!" he growled.

I looked around and could see a dozen black men glaring at me. I started to take off the pillowcase, then stopped. "Does anyone else know who I am?"

"No, I just saw you get out of your car and followed you in. I know you aren't like that, so do something. At least go to the latrine and fix it if you're going to stay here." He turned to leave and added, "I'll take care of the others and tell them you're just a dumbass who didn't realize what you looked like."

I hurried to my car and grabbed a Tennessee Volunteers ball cap from the trunk. Putting it over the pillowcase, it no longer looked questionable, and I went back in. Sensing I should straighten things out with Rich and his buddies, I ordered a round of beer sent to their table.

Once delivered, I went over and announced, "I'm just an asshole ghost, y'all."

Rich stared at me for a second and said, "It's OK, and thanks for the beers. Pull up a chair and join the rest of us f—king spooks." Their laughter broke the tension, and I hung out with them for the rest of the evening. Since Rich was from Chicago and his name is so common, I never managed to reestablish contact with him. He was one of our linemen on the 3rd SPG football team and a good friend. Despite that, he was a big fellow, and I wouldn't have wanted to tangle with him even if he was wasted drunk.

A friend of mine — Don Adams, with whom I would serve at my next assignment — told a great story on one of our NCOs. The NCO was a rather large man, and they often went downtown drinking together. One night at a bar, they took a couple of girls upstairs to a "short time" room. He described the rooms as no more than tiny cubicles separated by decorative curtains with grass and bamboo mats on the floors. While in the short time room, Don heard the girl with the NCO exclaim, "Hey GI, for someone so big, why you dick so small?" Don started laughing hysterically, whereupon the NCO tore down the curtain and started choking him. The guy let go only after Don promised to never tell anyone. Obviously, the story was too good to keep to himself.

The holiday season in 1981 was rough for me, and being off on weekends made the walls seem to close in. A lot of my single friends went home on environmental morale leave (EML) or took road trips. Getting mail from home didn't make me feel better, so I blew off steam working out and running the 3rd SPG obstacle course. Watching movies and throwing darts in the Peacekeeper's Pub got old after a while, so I preferred to stay busy. The chow hall put on great meals with lots of extras during the holidays, so there was always lots to eat. But I felt empty and alone, as if something was wrong with me.

While with SPAC, I had two inherent weaknesses — lack of experience (beyond my control) and immaturity (within my control) — that significantly hamstrung my effectiveness. Once my homesickness increased and became obvious, a meltdown was inevitable. Under normal circumstances, I would have been approved for EML to return stateside after a year on station. However, between designation as mission-critical and due to an unwed pregnant airman in another SPA section getting approved for CONUS leave, I was denied EML. This occurred in November, not long after football season ended. Following that, I tried to join a USO trip for a week in Beijing, China, during Christmas — also denied. I spent most of Christmas weekend working out, running, reading, or sitting on the barracks balcony alone, thinking and bored out of my mind.

For New Year's Eve, Don and I were requested to serve as a personal security detail for pop singer Imelda Papin's concert at the NCO Club. During a previous concert at Clark AB, she lost a very expensive diamond earring, possibly to someone snatching it during a photo hug. Our mission was to ensure no one grabbed her and to keep an eye on her jewelry. I waited at the Main Gate for several hours until her four-person entourage finally showed up in a nice car. They were very late arriving, so the motorcycle cops who were supposed to escort us finally drove away. So I escorted them in my Dodge Charger with my hazard lights on. I did manage to get the LE desk to stop traffic at several big intersections to ensure we didn't get delayed. We got to the NCO Club, and the Papin entourage ordered food, insisting that I eat dinner with them. Thinking modesty was appropriate, I ordered a cheeseburger with fries and a Coke. The Papin party looked at me quizzically, then (speaking in Tagalog to the waiter) placed orders too. I didn't realize the NCO Club was picking up the tab until after their steaks arrived. I felt dumb and thought I really should have inquired about that.

Don showed up before the show, and we stayed in contact via radio within the ballroom. We watched the concert and paid close attention to our VIP. Our effort paid off, and the singer wrapped up the concert without losing any jewelry or getting grabbed inappropriately. It was a nice evening, and Don ensured I had a couple of drinks at the bar despite my being in uniform.

Imelda Papin was extremely popular and, years later, served as the vice governor for her province.

The rest of the holidays were boring, other than a few blow-out parties on and off base. During one, I narrowly escaped a fight with the ex-husband of a female acquaintance from the honor guard. Apparently, he thought I was messing around with her. I wasn't, and only his best friend stepping in to confirm this prevented me from getting jumped on by a much larger man. I went to a couple of dances during the holidays, but something continued to not feel right, and I felt increasingly alone. My performance at work was starting to slip hard, and I also started pulling annoying pranks on others, including a couple of senior NCOs. Modesty aside, my artwork above the piss trough in the adjacent latrine shed belonged in an art gallery (wish I had taken photos). The latter infractions resulted in multiple verbal and two written reprimands. I didn't care, and it showed to even a casual observer. SMSGT Carrender pulled me aside and said, "Clark, I don't know what's eating at you, but you better figure it out and unf—k yourself most riki-tik."

SSgt Quesnell finally had his fill of my recklessness and referred me for a badly needed psych evaluation. Either I said all the wrong things to the doctors, or Don discreetly asked them to hold me for several days to decompress. I discovered it was the latter since the head psychiatrist (a full colonel with whom I remained in contact until his death decades later) declared me fit for duty the same day. Concurrently, he invited me — meaning that had I refused, he would have ordered it — to "hang out at our ward to decompress and sleep things off for a few days." So I spent five restful nights there and thus ended up receiving my twice-denied EML after all — even if it wasn't in Tennessee or China. While there, I saw the base dermatologist about a rash on my scalp and on both shins, plus I had a full physical with lab work. Incidentally, the shrink colonel told me quite a number of SPs spent decompression time there; one checked out the afternoon I checked in, then the morning I checked out, another SP NCO reported in. The place was almost like a merry-go-round for the 3rd SPG, a controlled environment for people to calm down. When I reported back to work following my "R&R," I felt like a million bucks and picked up where I left off. It's amazing what a few days in a stress-free environment can do, a tool I used successfully later in my career to help mitigate Soldiers' festering personal challenges.

While at the hospital, a friend brought me my mail. In it was a letter from my old friend, Hannah Hill, from Portland, TN. We met many years before at 4-H camp and could not stand one another at first. Over the years, we became close friends and always enjoyed one another's company. Hannah was attractive, athletic, and had an intense intellect. While always friendly to me, during high school, she lived too far away to date, so I never considered it. During our final 4-H trip to Knoxville, she tried to dissuade me from enlisting in hopes we could attend college together. I enlisted anyway, and we kept up a constant mail correspondence. This letter was different in that she expressed genuine love for me and wanted me to come visit her when I returned home in six months. It dawned on me that she might have had these feelings for years, and I was always either too busy or naïve to notice. I wrote her back, promising to come see her when I returned home. Our letter and photo exchanges were constant from then on and frequently crossed in the mail.

In January 1982, the base commander agreed to authorize SPAC to conduct a voluntary RSS at every residence in the Lily Hill housing and Hill housing areas. Lily Hill contained raised homes one could see underneath — built this way to prevent access by snakes and other varmints — and was predominantly housing for senior NCOs. Its streets were lined with huge acacia trees that provided near-constant shade throughout the year. One interesting feature was that the trees were painted white from ground level to five feet up. Although originally done to keep the long-gone cavalry horses from cribbing off the bark, the tradition remained. Following the completion of Lily Hill, we would move on to officers' housing up in the "The Hill" area.

With the base commander's support, Colonel Allison authorized SPAC to augment its manning by 10 more, with Don getting final say on selections. For several nights, I interviewed potential LE candidates during swings and mids at the chow hall. I gave their names to Don for vetting with their shift chiefs before he interviewed them. We soon had our 10 SPAC augmentees to train for our target hardening initiative. Among these was an older married airman named Oscar Reid from Abbeville, SC. I quickly bonded with Oscar and still regard him as one of the most reliable professionals with whom I have worked during my entire career.

For the next six weeks until late March, we conducted our target hardened area (THA) operations in these two large housing areas. We got the base engineers to replace street lights with newer, better lighting and remove random and unnecessary clumps of brush and banana trees to eliminate hiding places. We conducted RSSs on every home in each housing area and discovered deficiencies (a few serious, like electrical issues) for the engineers to repair. We taught the occupants a full range of physical security aspects (lights, locks, alarms, landscaping) and home security procedures (locking windows and doors, not showing off valuables, closing blinds, stopping mail when on leave, tearing personal information off items thrown in trash, etc.). We taught people to take proactive measures to defend their homes and persons against both common and area-specific criminal threats and activities. Remember, this was well before the era of ubiquitous cell phones and social media. We started early and ended late (typically after sundown), with our teams often holding information sessions with groups of 3-5 families on front porches. Of great value to all, we helped residents set up neighborhood watch programs that further organized some street blocks for social events — potluck dinners, babysitting, etc.

We made a lot of friends, and many wrote letters to the base paper supporting the initiative. Commanders took notice and wanted us to target-harden some of the off-base housing areas, so we arranged to do this after completing our base THA projects. We had conducted so many RSS in Carmenville that it was already an ad hoc THA neighborhood. Joining with Filipino landlords and local Metro Police, we conducted "train the trainer" events in two subdivisions. I believe these were Josefaville and part of the sprawling Diamond complex. The Angeles City mayor attended the former event and urged citizens to fight back against the criminals to make the streets safe for their children. The mayor even suggested that everyone should buy firearms for personal protection; I thought to myself, If you only knew. We were convinced of our project's benefit to all involved except the crooks we were helping to deter.

Best of all, both USAF and Filipinos alike were taking charge of their neighborhoods, thus freeing police to focus their energies toward interdicting other known criminal elements. Incidentally, SSgt Wallow and A1C Tinker were not involved in our projects. When first proposed, they reportedly voiced opposition, so the 3rd SPG leaders purposely omitted them thereafter. I sensed trouble brewing from their increasing verbal taunts. Sgt Labunski (star of a previous crime awareness TV ad on Far East Network) worked in Pass and ID and warned me he had overheard them talking trash about our projects to new arrivals in the SPG HQ. He said, surprisingly, a few other NCOs seemed to share their sentiments. He cautioned me, "Watch your back and don't trust them." I already knew this, but the warning meant trouble was festering.

One night, Oscar and I stayed long after dark in the Lily Hill NCO housing area to assess exterior lighting recommendations we made during RSS visits that week. He had picked me up that morning, so he dropped me off at my barracks en route home around midnight. At some point during the night, someone jimmied the door to SMSgt Carrender's office with a sharp object, then the perpetrator used the chief's blowgun to shoot darts all over his office and generally wreck the place. Adding to the intrigue was that the formerly pregnant airman mentioned earlier (now with a baby at home) reported seeing me around 2300 that night, exiting the building as she allegedly drove past from a friend's home. Seriously, what was she doing out at 2300 when she had to report to work 8 hours later? And where was her baby during this time?

The next morning, I reported in at 0730, and Don asked me where I was the following evening. I told him and said Oscar could confirm it when he arrived, adding that multiple LE patrols from the swing and mid shifts were present in Lily Hill until we left and could verify our whereabouts until almost midnight. Oscar soon arrived, and Don talked to him alone to confirm our location and timeline. Having jotted down notes as to the NCOs' homes we had visited, Don called two at their units. Both confirmed Oscar and I left Lily Hill housing around midnight. Don then asked, "If you were going to break into SMSgt Carrender's office, how would you do it?" While I found this a strange question and told him so, he said, "Indulge me." I replied since every office in the building shared a common attic above the drop ceiling tiles, anyone in the building could get on a chair, move one tile, climb up, and then access any office they desired through the attic. He then told me what occurred the previous evening, that I was accused as the culprit. "You're going to get questioned, so don't take it personally. Stay calm." Don thanked me for my input, and I went to my desk to prepare for the day's RSS agenda.

Suddenly, an admin TSgt lunged cigar first through our door and into my personal space.

"Chief wants to see you in his office right now."

I said I had to first lock up a couple of documents.

He growled, "No, right now! Get moving, Airman!"

Obviously marking time until retirement in a couple of months, this loser's wheel was still turning, but the poor hamster was long dead.

Carrender stood in his doorway and looked mad enough to chew railroad spikes. He pointed at his upended office furniture and spilled binders and asked, "Did you do this shit?"

"Negative, Chief." I could tell he didn't believe me. "I was in Lily Hill NCO housing until almost midnight, then Airman Reid dropped me off at my barracks. I can provide a dozen witnesses to prove this, too. Perhaps the guilty dogs have already barked with this sad setup attempt and their groundless accusations."

He snapped, "Don't get smart. But what do you mean by that?" I now understood why Don asked me how I would have done it.

"Look at the signs of forced entry. As many times as you've sent me on errands, don't you think I would have made impressions or copies of your keys if I had ill intentions toward you? In fact, keys to the SPG headquarters and all SPA offices are on your key ring. Beyond that, it defies logic to enter the SPA building with no signs of forced entry, then jimmy your door, leaving evidence, and then trash your office. Your door opens inward, so slipping in a thin piece of hard plastic or scrap tin will pop it open in seconds and leave no evidence. Obviously this was done by some very stupid asshole trying to discredit our crime prevention initiatives." I stared directly at multiple witnesses who stood several feet away, eavesdropping on our conversation. I asked Carrender if he wanted my response in a written statement, adding that Airman Reid could also provide one. He shook his head that he didn't, and said I could go.

Satisfied it was a setup job, SSgt Quesnell was very upset over the accusations against me. He reported his findings to SMSGT Carrender, who still didn't believe us. Don then said, "Well, if Dan had done it, he would have entered through your ceiling tiles to avoid signs of forced entry. Or he would have slipped in a plastic card at the strike plate. And if he really wanted to get to you, he could have just burned the building to the ground." I heard him offering the evidence he had collected that would exonerate me, but Carrender declined. We suspected he had already connected the dots and figured out the culprit(s), so now we wanted this saga to go away.

Within an hour, several people walked by the office door to make their suspicions known, insisting I was the guilty party. I advised them to go get bent. One told me to watch my mouth — as if his rank meant anything now — so I stood up and asked if he really wanted to dance. He knew I would crush him like a bug, so he just sneered and walked away. I thought, These stupid bastards don't know I have a loaded .45 in my briefcase.

My prankish chickens had come home to roost despite clear evidence of my innocence in this one. Worse, my paranoia instantly surfaced, and I decided my best option was returning to LE duty. Whereas I reasoned that if someone wanted my head, they'd eventually take it, being legally armed during duty hours was my best defense. With great regret, I requested that SSgt Quesnell have me reassigned to A Flight LE once our THA projects were complete. He concurred and added, "I think I'm probably right behind you." Don was also done with the drama and ready to transfer his family to CONUS.

I had toyed with the idea of taking a consecutive overseas tour (COT) to South Korea or extending at Clark AB. Several Southerners had received follow-on assignments to northern tier SAC bases like Minot or Grand Forks in North Dakota or other places where winters were awful. Those I had heard from were miserable and urged me to extend at Clark or take a COT to Korea, then reenlist to return to Clark or a better stateside base. I sensed merit in their warnings and felt it prudent to consider all options. During my only phone call home in early February 1982, I discussed it with my parents. They preferred I return stateside and hope for an assignment near enough to drive home on leave, plus with so many bases in the Southeastern US, I had a high probability of getting one. When I hung up, I drove out to an area east of the flightline with a couple of San Miguels to think it over. I pulled up next to a derelict aircraft that once belonged to the vice president of South Vietnam and watched the sun set over Mount Pinatubo. The spectacular sunset over the Zambales Mountains helped me focus as I weighed the pros and cons.

I decided I would forego the COT tour or extension request and take a chance on getting an assignment near home.

While working the target-hardening project, I received orders to Barksdale AFB in Shreveport, Louisiana. Simultaneously John Marino received orders to Carswell AFB in Fort Worth. John was about to marry another airman and would extend to 36 months at Clark. Agreeing that Carswell was a better location for me, we went to the base personnel office and requested an assignment swap. Although Fort Worth was an 11-hour drive and mostly meant flying rather than driving the 8 hours from Shreveport, it was a bigger city with much more to see and do. I also knew several SPs who had served at Carswell and they all enjoyed their tours. Two weeks later we received our new PCS orders for our respective bases. Everyone was happy and I think my parents were the most happy with the orders.

In late March, I rejoined Alpha Flight as a patrolman. I worked a nine-day shift — six days on and three off. Due to promotions, transfers stateside, and several demotions by LEs behaving badly, I was now the most senior E3 on Alpha and due for promotion in six months. MSgt Breeze Halbert was still around and was glad to have my SPAC experience on board. I was surprised when he told me I would work mostly patrols and minimal gates. He wanted me to help train several new NCOs and young airmen just arriving from the States. I asked if he had any problem with me officially joining the EST team, and he was fine with it. Although not officially an EST member, MSgt Funk had pulled me for quick reaction force (QRF) duty as an M60 gunner during several exercises. Strangely, I wasn't asked to try out, and after I mentioned wanting to join, they added me to the EST roster.

The same month, we had a big exercise, and as usual, the EST team pulled QRF duty. We deployed a few times on simulated targets, but mostly waited under shade trees in our Duck armored vehicle, listening to the radio and waiting for calls. I did pushups, sit-ups, and calisthenics to pass the time. Other team members played cards or stood watch. It was a great duty compared to walking the flightline in the sun, getting blasted by jet wash, and breathing fumes. Although we received a few passing catcalls about shamming, no one ever said it directly to our faces. We had some real studs on EST who were more than willing to settle any perceived duty grievances with fists.

Our Duck was a Vietnam-era armored vehicle with four large wheels, bullet-proof driver windows in front, and small circular firing ports along the sides. It also had a machine gun mount on the top accessible by steel doors that folded up and sideways. Back stateside, the Air Force had already begun to replace its Ducks and M151 jeeps with Peacekeeper vehicles (specially designed for flightline use) and Humvees. By the mid-1980s, the transition was complete, and all Ducks seen today are static displays at museums or in private collections. The 3rd SPG's M113 armored vehicle stayed in the motor pool because its tracks tore up the paved roads. The only time I saw M113s in operation was at Camp O'Donnell during exercises.

Having worked in Crime Prevention, where I was trained to analyze incidents and trends, patrol duty was rather easy. The older NCOs openly trusted my judgment and were more than happy to let me train the FNGs as an area patrol leader. I taught them how to read terrain and spot concealment that intruders could use for infiltration and hiding, plus how to use cutoff points to interdict escape routes. I made sure they knew where storm ditches and washout drains crossed the perimeter fences, as these were common escape routes. I also taught them how to read potential vulnerabilities in base housing areas and point these out to people living there. Doing this restored my sense of purpose, and I felt I was again making a difference. Although not yet 20, my older and younger peers seemed to defer to my experience.

SSgt Quesnell soon joined Alpha Flight and was designated a

patrol supervisor. We worked my same nine-day shift cycle (6 days on / 3 days off), so he was welcome company. He was disenchanted with what had transpired in SPAC before returning to road duty and disgusted with the 3rd SPG leadership for letting it happen. We heard Wallow and his new crew were steadily dismantling Don's crime prevention projects in favor of more passive, easier to manage programs — superficial public-relations gimmickry that sounded great and briefed well, yet never produced measurable results. Concurrently, word on the street said the crooks and communists quietly celebrated, and over the next few years this greatly increased the risk of violent crimes against both US personnel and Filipino civilians working on the base. With Don Quesnell now out of the picture, seeing happy unicorns farting confetti fly out of the "new and improved" SPAC office would not have surprised anyone aware of the saga.

On a more positive note, Belle Quesnell gave birth to Maria that May. Other than my brother Shaen and cousin Chris, I had never really wanted to hold a baby before. Holding Maria was different, and I regretted not spending more time with the Quesnells. Many years later, while visiting them in San Antonio, I scooped up the teenage Maria and said, "Last time I did this, you were wearing diapers." Thinking it funny, Don snapped a photo. Besides serving as my mentor and longest supervisor to date, Don was the big brother I wish I had had growing up. His mentoring and friendship made me a better human being.

Another positive occurred around 1 April when my friend Christine "The Valkyrie" Debevec arrived in the Philippines. We met while I was in SP School in late 1980 and hit it off, so we stayed in touch. She had a serious boyfriend from her previous assignment at Grand Forks, North Dakota, and he was trying to get assigned to the Pararescue unit at Clark AB. Christine and I hung out together some over the next few months, and the more I found out about this significant other, the more I sensed it was a toxic relationship. He allegedly cheated on her numerous times and was abusive, yet she was adamant that she could make it work. Fast forward several years, the ensuing marriage was a disaster and didn't work out. I abruptly lost contact with Christine in the late 1980s and, for years, occasionally tried to track her whereabouts. In 2021, I finally discovered via a mutual friend why she disappeared. She was killed in a head-on collision while driving cross-country during leave. Despite the four decades that had passed, finding this out about my friend greatly upset me.

One day, while on an LE patrol, the desk called out to ask if anyone knew how to rope a cow. I answered that I did and was directed to meet the Animal Control team in Lily Hill base housing. Somehow, a 1000-pound steer had gotten out of a cattle trailer on MacArthur Highway between Mabalacat and Dau; it entered the base through one of the many holes in the perimeter fence. The transport team chased it across the flight line, where it turned west and passed through Barracks Row and the streets between the Transportation and Supply complexes. By the time I arrived, the steer had turned south and circled behind Kelly Theater and the bowling alley. On a hot day like that, running cattle can make them overheat to the point of death. I kept thinking, Where is Felix? He knows how to rope cattle. Turned out he was on leave or something, so he wasn't available. We had a lot of country boys, yet no one answered the radio call.

So it fell to me.

I finally caught up with the steer as it crossed the open field northwest of the chow hall. It trotted across the bridge over the deep drainage ditch and almost to the chow hall. It finally moved into the shade behind the adjacent barracks; I could tell it was thirsty and looking for water. Multiple cars were following along and actually made the situation worse by further scaring the beleaguered bovine.

Two tired-looking Animal Control guys greeted me, and one handed me a rope. He said, "Hi, Sarge. Can you rope him?" I looked at the equally tired steer and said that I could, but wanted the growing crowd of curious onlookers to get back at least 50 feet. Knowing the steer would stand still while drinking water, I yelled for someone to get two big pots of water from the chow hall. Meanwhile, I fashioned an expedient halter from the rope.

Minutes later, I had the water and eased it in front of the steer. It began gulping down water, and I stroked its back and neck to help calm it down. When it lifted its head, I slipped the halter over its nose and ears, then let it drink the second pot of water. The steer was calming, and I had one of the Pinoys hold the rope while I used another rope to tie a 2-foot hobble onto its back legs. It never moved while I did this, and the steer was now under my control. I called in to the desk that the package was ready for pickup.

Onlookers were crowding in again, and many said they had never seen a cow this close. I kept correcting them that it was a steer since its balls were removed. For about 15 minutes, I held court with the ignorant and uneducated about cattle nomenclature and bovine anatomy. I was glad when the truck hauling the cattle trailer finally appeared. I removed the hobble, and they led their lost steer to the truck for loading.

By now, Colonel Allison had heard about the wild steer chase across the flight line and into the cantonment area. He showed up and asked, "Clark, are you the rodeo clown who roped that steer?" I replied, all I did was get it water so I could put a halter on it, that people nearly gave it a heat stroke chasing it across the base. He laughed and said, "Stop by the Peacekeeper's Pub after your shift ends. I'm coming over to buy your crazy ass a beer." True to his word, Lone Star stopped by with Chief Fields, and we had several beers. Both were from Texas and liked cowboy stories, despite this one being a rather lame tale.

Several EST guys were assigned dual duties as designated marksmen. "Sniper" is a way overused, inaccurate term since our job was more to deter threats than eliminate them. We took a short marksman orientation course run by the Marines at Subic Bay Naval Station to get certified. I had high overall scores and also did well with a Starlight scope. Afterwards, we trained at a firing range nestled in the foothills beyond the POL dump. Having plenty of trained marksmen allowed their use for engagements in high-risk incidents and as security in the weapons storage area (WSA) on the east side of the flightline during alerts.

Although it wasn't discussed (much less admitted by the DoD), everyone believed there were tactical nukes in the WSA. The triple fence, imposing central guard tower, and tight security spoke volumes about what might lie inside those igloos. The WSA was surrounded by a couple of hundred acres of grass kept bush-hogged down for clear fields of fire. While the area lights were usually dimmed, these could brilliantly illuminate and blind everything within and far beyond the fences. Deadly force signs in multiple languages every 50 feet around the outer fence were another stark reminder to remain clear.

During one alert, I was assigned marksman duty in a tower, an easy and boring gig. A Marine Gunnery Sergeant ("Gunny") was in charge of the tower and rarely said much as he monitored a panel of sensors. Occasionally, he went outside the door to smoke a cigarette and stare off past the lights pointing out toward the fences. At night the lights were normally kept very dim both to preserve night vision and to prevent blinding a response force until it could deploy. I pretty much sat on my rump and watched the fences or clouds off in the distance to pass the time. I spit at a flat rock down below and had it half covered before the Gunny got annoyed and told me to stop. Again, working in any bomb dump was a boring duty and I was glad when my relief arrived.

One day, I was patrolling the housing sector west of 13th Air Force HQ in an M-151 jeep when several monkeys darted across the road right in front of me. I struck one and felt the thump-thump as both wheels crushed its body. The NCO riding with me howled with laughter and made monkey sounds, so I expected more pestering over this accidental yet righteous kill. This wasn't worth reporting to the LE desk, and I thought nothing more of it despite a few snickering clowns making chimp sounds during guard mount the next day.

Several days later, Chief Fields called me over after a shift ended. He grinned as he handed me a wooden pointer with a dried monkey paw on the end. "Take this mojo wand home as a souvenir." One could buy these downtown, so it clearly wasn't the one I ran over. Unfortunately, US Customs seized it from my carry-on baggage when I arrived in Los Angeles on my way home. I'm sure some dickhead Customs agent kept my mojo wand as a trophy!

A bunch of us were eating at the Pizza Factory one night, and I kept seeing something move along the floor and then dart behind the juke box. I fixated on it and soon noticed it was a mouse. Someone observed that people frequently fed crumbs to it, hence its frequent forays into the danger zone of patrons' feet. A half-drunk FNG whose name I've long forgotten took a small piece of pizza crust and dropped it next to the wall. When the mouse came out and began feeding, the guy nailed it with an empty San Miguel beer bottle that shattered everywhere. The mouse died instantly, and the Pinoy manager made the offender clean up the mess. Thereafter, I only knew the young troop as "Mouse Man."

Now among the oldest E3s in the 3rd SPG and with my E4 promotion date approaching, I was invited to cookouts and dinners by the mid-grade NCOs. Usually, after their wives fed us awesome Pinoy food, we talked and played horseshoes or cards. Of course, there was frequently a drinking game or two. My favorite was called "The Beer Hunter" — inspired by the Russian roulette scene from the movie, "The Deer Hunter." The host passed around one beer from every six-pack of cans, and players took turns shaking it up; the host then placed the shaken cans in the cooler, and every participant rearranged the cans to ensure no one knew where the "bullets" were.

Each man took a beer and held it to his ear to crack open. Whereas non-shaken beers just popped open, shaken ones usually exploded all over the holder's ear and head. Obviously, the game's object was to not blow beer all over yourself. While I somehow never managed to pull a "bullet" from the coolers, others were not so lucky. One young NCO had a nasty evening by picking three consecutive "bullets" from the same cooler, much to the amusement of his peers and their wives. Such off-duty games were highly entertaining. Years into the future, I made sure our fraternity pledges learned how to play "The Beer Hunter."

One day while on Negrito Gate, a breeze was blowing a horrid smell from the nearby dumpster. I checked inside but could only see rotting garbage covered in hordes of flies. The gate shack kids said someone had thrown in several large bags of spoiled food a couple of days earlier. It stank to high heaven and was definitely a health hazard due to the flies.

During a gate check around midday, LT Luena stopped by and complained about the smell. He directed me to report it to the base engineers; I replied that, according to the guard ledger, it had already been reported several times. He didn't like my answer, so he told me to find a way to get that dumpster emptied out since it was a health hazard. I thought to myself, Aren't you the officer with all the authority? He went on to observe that we had also tracked dirt onto the gate shack floor, and my boots were dusty. Never mind that this gate shack sat on a large patch of bare dirt and volcanic ash. While I didn't dislike him personally, this really annoyed me. Of note, many years later, as a lieutenant colonel, I looked him up online and made contact to thank him for his leadership lessons without specifying that some were the ones I ensured I never emulated.

After the lieutenant left, I fumed over his nitpicking. I called the LE Desk, and they already knew about the dumpster. So during chow relief, I asked the patrol leader to swing by the mini-BX next to our barracks so I could run in and buy two things. When I returned with two cans of lighter fluid and a book of matches, he laughed and said, "I know what that's for and Luena's gonna have your ass over it." I told him I'd take my chances. He dropped me off, still chuckling. Once his jeep was out of sight, I walked over to the dumpster. Fighting the flies and gagging from the stench, I squirted both cans of lighter fluid into the dumpster and then set that sucker on fire.

Within two minutes, a massive column of thick black smoke was pouring out of the dumpster. Obviously it contained either rubber or plastic items underneath the rotten food bags because the fire grew intense in a hurry. I returned to the gate shack, where the PAF guards pointed and spoke excitedly in Tagalog. Suddenly we heard muffled but still loud explosions, as only God knew what was detonating from deep inside. The fire grew so hot that the exterior paint began cracking and flaking off the dumpster. Watching the smoke billow skyward, it dawned on me that probably the entire base could now see it. By now the PAF guards and all the gate shack kids were laughing. One PAF guard exclaimed, "Sarge, you f—k up reeeeeal baaaaaad! Hey, Lootenant, coming!" I looked down the long road to see LT Luena's patrol car heading my way.

Uh-oh, I'm in deep dookie now.

As the car stopped in front of the gate shack, the kids disappeared and the two PAF guards acted as if nothing had happened. Another patrol vehicle approached right behind his, and I was relieved to see it was Chief Fields. As the chief emerged from his vehicle, Luena was already out and addressed me by my rank in a calm but firm voice. "Airman First Class Clark, why is the dumpster on fire?"

Seeing Chief Fields grinning at me, I thought I might actually bullshit my way out of this goof-up. Motioning toward the PAF guards, I replied, "First Lieutenant Luena, sir, I believe — as do my esteemed Philippine Air Force colleagues — that this dumpster fire is directly attributable to spontaneous combustion."

Luena narrowed his eyes in clearly annoyed disbelief and asked, "How so?"

"Sir, the kids here said someone threw in several bags of rotten food a few days ago. When wet items are thrown on top of combustible items, then additional combustible items are added on top of them, the pressure of the weight, coupled with the internal heating from the sun beating down on the metal dumpster, can create extremely high heat in such a tightly confined space. I contend it eventually grew so hot that the whole damned thing just burst wide open into flames. Plus, my PAF friends here can attest to multiple loud explosions as items inside have indeed detonated. We think someone illegally discarded flammable items inside, hence the black smoke. You can ask them if you like, sir."

I turned to look at the PAFs, and they nodded their heads a lot as LT Luena spoke to them in Tagalog. I understood enough Tagalog to know they wholeheartedly agreed with me. The lieutenant looked at Chief Fields, who remarked in a serious tone, "Sir, that sounds completely plausible to me. We have several more stops to make before the shift change. If you don't mind, why don't you handle those, and I'll address the situation here with these shithead gate guards. Sound like a plan, sir?" Luena nodded his head and I saluted him as he got in his car. He never rendered my salute nor looked back at me as he drove away. The PAFs skedaddled inside the gate shack and laughed like circus freaks.

I turned back around and Chief Fields had walked up to within a foot of me. I snapped to attention, and he snarled, "Damn you, Clark. You need your ass kicked. Why do you keep doing stupid shit like this?" The PAF guards were snickering so loudly behind me that the chief then began laughing. "Geez, you should go into comedy. Personally, I couldn't care less about that dumpster. You were told to do something about it and took some initiative to eliminate the problem. But no more fires, and I mean none. I know you don't smoke, so I'd better not catch you with any matches or lighters either. Do you understand me, boy?" I replied that I did and thanked him for his leadership.

As he got into his vehicle, he turned and said, "For someone everyone swears is so smart, you really are a stupid son of a bitch sometimes. And if anyone asks, you'd best stick to your story. If I hear you've bragged about defying the lieutenant, I'll put my boot so far up your ass you'll taste the leather and have a black tongue from the polish." He stared at me for a moment and added, "You better be glad I see something in you." I saluted him as I always did and he sharply rendered it. Of all the NCOs with whom I served over my three decades in uniform, I respected and loved Chief Fields the most.

I found it odd that the base fire department never drove out to extinguish the fire. After a while it burned out and the smoke cleared enough to slide the side doors back and look inside. We discovered the cause of all those muffled detonations: several dozen paint cans. Of course, while turning in weapons after our shift ended, word already got around that I torched the dumpster at Negrito Gate. When peers asked me what happened, I responded that Chief Fields and I believed it was spontaneous combustion. Standing near the clearing barrels, the chief looked at me with disdain and jokingly shook his fist at me as I walked by to my car.

Chief Fields soon returned to SPA and our new flight chief, SMSgt Lippi, was a wonderful fellow to work with. He assigned me mostly to patrols and I enjoyed them. I particularly liked walking patrols and getting off work 60-90 minutes early. Electrical power was rationed for two hours daily and cut to our section of Barracks Row from 1600-1800 as part of the rolling power blackout. Several barracks friends had blenders, and whoever got off work first would rush back to make daiquiris and margaritas; that individual bought 3-4 bags of ice at the Pizza Factory, threw in fruit and rum, and started blending. We had a couple of small plastic garbage cans that we used for nothing but frozen drinks, so we blended until they were full or power was cut at 1600. This way, we had plenty of drinks for the rest of our buddies getting off shift. We sipped on these until they ran out, left to eat chow, then returned to blend more frozen drinks once the power came back on at 1800.

One day, while patrolling the Mabalacat sector in my jeep, I was called to the picnic ground near the stables to run off some kids throwing rocks at the picnickers. A new E4, SrA Petty, was riding with me, and this was his first chase call. Upon seeing our approach, they took off into a narrow wooded area, through a ravine, and into the elephant grass behind it. With our jeep not yet stopped, Petty jumped out and chased the kids into the elephant grass. He quickly disappeared, and I thought, Oh crap — ambush! Drawing my pistol and running up to the edge of the grass, I yelled, "Petty, get your ass back here!" I stood there aiming into the grass wall for a couple of minutes, and he reappeared. He asked me what was wrong, and I snarled, "For all you knew, we were getting drawn into an ambush, and those kids are gone. Never go into the grass alone! That's how people get killed here!" We remounted and rode in silence for a while. Petty finally spoke up and said he was fine with the correction. I apologized anyway for my tone, and he said he didn't take it as disrespect. I liked Petty because he understood the Philippines was unlike any other place he had been stationed before, and he was willing to learn the ropes.

In late spring, right before the monsoon started, Pinoy workers started clearing the underbrush from the north side of Lily Hill. The cut brush lay drying in the sun and was soon perfect for a good burn. One day, while patrolling alone, I pulled behind the long-abandoned Missile Maintenance Squadron barracks, hiked into the mess, and set it on fire. I drove away as if nothing had happened and returned to my patrol sector. It didn't catch well and quickly burned itself out. The next day I tried again — same result. A few days later while riding with a trusted friend, I did it right. Driving to the top of Lily Hill, I walked down the north side 100 feet and set a good one. We drove over to FM Hill near NCO housing and my friend took a photo of me with Lily Hill smoking in the far distance. The brush was too green to spread beyond a 30-yard circle, so the fire department let it burn itself out. The monsoon was upon us and the rains ended my pyromaniacal aspirations for good.

Around this time, other Filipino workers started hacking down the tall brush (banana trees and bamboo) east of the big bomb dump on Mabalacat Highway. I found it sort of amusing since it would all grow back in a month or two. Other workers replaced the bomb dump fence that thieves had incrementally cut down during the preceding months. I remembered what Breeze Halbert had told me a year earlier and took photos of a long stretch of the fence. Within two weeks, holes started appearing in this new fence, too. Ultimately, thieves eventually stole anything that wasn't kept under constant observation.

One afternoon, after I turned in my pistol and was leaving the armory, I was told to stand by for a special mission. An older NCO said we were picking up an airman dealing drugs out of the BX restroom. The sergeant laid out a simple game plan. "We go in quietly, radios off. This guy deals out of the middle stall and responds to a series of two double knocks before dealing underneath the wall. I'll go into the stall nearest the wall and you" — pointing at my buddy — "watch the latrine door and not let anyone else in." Pointing at me, he said, "When I yell 'GO,' you kick his stall door in on him. I'll come over the top and use my nightstick on him if he resists. You do the same. We'll cuff the clown once we have him under control. That's the plan — any questions?" I thought, Wow, this is a new twist on dealing 'shit' but sounds like fun.

En route to the BX, a radio call came that confirmed our mark was in place with no one standing watch for him. We pulled into the BX and tiptoed into the latrine. Sure enough, a pair of boots was visible underneath the door of the middle stall. The NCO nodded to us and then hurried inside the stall nearest the wall and closed the door. I could hear a few faint taps that were answered by a few more. I saw a hand offer some money under the wall, and another hand offering a bag going the other way. The sergeant yelled, "GO!"

Pulling my nightstick, I reared back and kicked the door in on the dealer as the NCO came over the top. The skinny young airman threw up his hands in surrender too late to stop the NCO's club from crashing down on his shoulder. As I grabbed the guy and slammed him to the floor, the NCO pounced down to cuff him. Through our yelled orders, the guy shrieked, "I give up, dammit! I give up!" I'm not sure whether he understood the rights warning that followed, but he definitely understood he was under arrest.

Our quarry and evidence bagged, we walked the druggie past gawking shoppers for the grand walk of shame out to our jeep. The

NCO crowed, "Y'all take a good look at this punk. Deal drugs and you'll go to jail." We then headed to the LE desk to turn the airman and evidence over to the desk sergeant, plus complete our witness statements. Afterward, we went to the Peacekeepers Pub for a celebratory beer.

One afternoon while working the Main Gate pedestrian lane, I saw a Flying Tiger jet roar off to the south and circle high to the east and then north, climbing onward to take others back to "The World." I stared as its blue tail fin melted into the sky, leaving only its white fuselage and wings in sight; the 747 diminished to a faraway point of light and finally disappeared. Watching the plane made me feel homesick. Observing this, a PAF airman working with me walked over and patted me on the shoulder. He said, "Sarge, soon you go home to your family." I got along rather well with the PAF guards and knew many by name. I frequently bought soft drinks, brought chow hall fruit, or doled out my cigarette ration to them, so I got the impression that I was popular to work with. Toward the end of my tour, I understood a lot of Tagalog and could converse with them some, plus knew all the bad words, slang, and insults. I learned key phrases and could ask for IDs and such in their native tongue. I found that when I started a conversation in Tagalog, the listeners usually shifted to English. It was unwise to speak badly in the presence of one who might understand.

When working the Main Gate, my favorite morning treat was pandesal, a semi-salty yeast roll. A medium-sized bag cost about 20 pesos (two bucks US) at the bakery across the taxi parking lot. If I got posted early enough, I could send a PAF guard to buy us a big bag and some coffee or soft drinks. Within minutes, we had hot bread and drinks. With 6-8 guards (2 USAF and 4-6 PAF) working the Main Gate complex, the pandesal never went far.

Town Patrol HQ was just outside our pedestrian lane, and I'll never forget the carved wooden sign above the inside of the front door: "We are the Unthanked doing the Unnecessary for the Ungrateful." The extended version continued: "We've done so much with so little for so long that we're now qualified to do anything with nothing." A friend who graduated from LE School two classes behind me (Ron Payne) worked at Town Patrol. He liked to harass me whenever I showed up there to pee, so as retribution, I took trophy-sized dumps and left without flushing. Ron and I reconnected years later on social media, and he still busts my chops about everything possible.

198206: A1C Matt Rossoni stands with a PAF guard and local kid at Negrito Gate in June 1982. A former police officer in California, Matt made the Air Force a career and retired as a Senior Master Sergeant.

Town Patrol was officially known as the Quad Agency Patrol due to the four agencies represented: USAF, PAF, Angeles Metropolitan District Command Police (AMDC), and Philippine Constabulary (PC). They worked a large area, so we rarely saw them unless they stopped at the HQ or entered the base. Thinking it a glamorous duty, a lot of airmen wanted to work on Town Patrol. In reality, the duty was difficult and hazardous according to those who worked there. They had to investigate the full range of criminal activities and deal with bad characters who would have cut their hearts out if given the opportunity. They investigated shootings, robberies, rapes, and assaults of every conceivable type, often escalated by the booze that flowed freely in the bar districts along MacArthur Highway and Fields Avenue.

The area was full of US expatriates, and some were interesting characters. Among them was a USAF veteran named Philander Rodman, father of future basketball star Dennis Rodman. Philander owned a bunch of bars and had a large number of women working for him. He was known for flaunting this to authorities, many of whom he likely paid off. While I met him a couple of times and never had any negative encounters with him, many SPs did over the years. One time when Town Patrol arrested him for something, he set his clothes on fire in the drunk tank and then put out the fire by pissing on them. He was hauled before the night court judge wearing scorched, urine-reeking clothes, so the judge was not impressed. I don't recall the outcome, but due to stories like this, I avoided him; he was also rumored to have sired more than 50 children. It was years in the future before his estranged son Dennis became a famous pro athlete.

Another day working the Main Gate's outbound lane, I had to inspect a large cargo truck heading to Manila. I looked at the cargo manifest, and the TO line read "JUSMAG - Manila." I knew this was the Joint US Military Assistance Group, a significant DOD activity that handled all direct US military support to the Philippines. This included logistics, weapons procurement, training assistance, meteorology, and exercises, plus coordination with Pacific Command and other allied countries in the region. When they opened the big door and I climbed into the back, I was surprised to discover that about half the cargo was whiskey, liquors, and fine wines — many dozens of cases. I also noticed a big paper grocery bag with Playboy and other American porno magazines spilling out. I asked the American civilian riding shotgun why there was no rum, and he shrugged. Rum is plentiful here. I just transport the stuff that's hard to get." I waved the truck through, and the Pinoy driver saluted me. Only in the Philippines.

The PAF HQ was a couple of hundred yards from my barracks. Like a couple of other PAF buildings, it had numerous fruit trees around it. The PAF HQ had a bunch of papaya trees loaded with fruit, and they always waited to pick them when ripe. One night, walking back from the chow hall past the PAF HQ, I somehow thought it was a good idea to go up on the porch to check the papayas. As I neared the trees, a very drunk PAF MSgt named Garza stepped out of a room, pulled his .45, and screamed, "Sigi na, putang!" Knowing he meant "get the hell out of here, you whore" I jumped off the porch and sprinted for the street. A few days later, I saw Garza and asked why he drew his pistol on me. Looking astonished, he put his arm around my shoulder and replied, "That was you, Sarge? Next time, bring San Miguel and we'll drink, OK?" Apology aside, I steered clear of the PAF barracks for the rest of my tour.

When patrolling in those sectors, I made it a point to check the PAF guards in the towers around the NCO and officer housing complexes on the northwest end of the base. Per the Bases Agreement, the PAF was supposed to man every tower; in reality, it manned few, and the 3rd SPG ended up posting DOD guards in critical areas. Once I checked a PAF guard and discovered he had only four rounds of ammo for his M16. These guys were no match for well-armed criminals or insurgents who decided to test them. However, a year or so before I arrived, a PAF guard shot and killed an intruder near the elementary school, thus the bad guys knew the risk before coming over the wall.

Some housing area residents pooled their money and hired private guards. Nearly all were Aytas (Negritos), and the crooks avoided them. Their small size, dark complexion, and agility made the Negritos deadly foes in the darkness, and their reputation as mountain-style insurgents was legendary. While we heard plenty of stories about Ayta guards killing thieves and intruders at night, I knew of only one confirmed incident during my tour. That was the morning I had to go stand guard over the headless corpse of an intruder in officer housing.

Once the monsoon rains started, we usually saw the sun at sunrise and sunset because it was typically cloudy or rained most of the day. Whenever it emerged at midday, the temperature rose quickly and generated billowing clouds and rain. The first summer I was there, it rained nonstop for a three-week stretch, followed by a couple of mostly sunny, horridly humid days and then three more weeks of solid rain. While it occasionally rained during non-monsoon months, we could expect rain every day from 1 June - 1 October, with the peak around 1 August.

The sunrises over Mount Arayat were sometimes breathtaking, including this one in April 1982.

I recall times when the large stormwater ditches overflowed so much that they resembled creeks. As buses swung off the main roads to the bus stops, their weight and tire angles created large mud holes in front of the awning-covered benches. Early each morning, airmen sneaked their female guests out of their barracks and sheepishly waited for buses to pick them up for transport to the Main Gate. It was obvious they had broken the base regulation prohibiting overnight guests. On weekend mornings, this was prevalent, and sometimes after a daybreak run, I went hunting for frenemies and sports foes to splash on the way to the chow hall. I usually took a runner friend or two with me to refute any complaints. If I drove by a bus stop and saw someone I didn't like — honestly, "like" wasn't a protective criterion — I drove around the block and returned. Then, before passing that bus stop, I gunned the engine and swerved hard into the deep mud holes to send a curtain of water and mud careening onto them and their female guests. This ensured they received at least some sort of bath after their presumptive night of carnal adventures. However, if I saw PAF or other adult Pinoy workers at the early morning bus stops, I stayed on the pavement and drove on. Whereas I knew that airmen breaking regulations would never report me, the PAF and base workers would. While I was a real dick at times, I respected my work colleagues and our Filipino civilian workers.

One very rainy Sunday morning, not long before I transferred out, two running buddies and I ran past two airmen from a maintenance squadron with their overnight guests sitting at the bus stop by the PAF barracks. They were all alone, and I had an unresolved grudge against this one clown from a football game the previous fall. We were beating the hell out of his team, and the guy elbowed me in the mouth, causing me to leave the game until the bleeding stopped; I remembered the bitter aluminum stick burning my lip and gums to stop it. As we reached our barracks a block away, I asked my running buddies, "Y'all got time to fight some deserving assholes this morning?" Both said they did so, we jumped in my Charger and drove straight back. I splashed the cretins thoroughly, and we could hear them yelling and cursing as we left.

We drove on to the Main Gate, and I was happy to see the day shift had just assumed post. We parked by the mini-BX, and as we exited the pedestrian lane, I asked the A flight LE and PAF guards to not respond to what might transpire. I told the PAF NCO I was about to "beat up an American bakla," and he just laughed. He yelled, "Kick his ass, Sarge!"

I asked my buddies to jump in only if the guy's buddy tried to help him because I wanted to keep this fight honest and fair. Sure enough, when the bus arrived, the two maintenance guys followed their guests outside the gate. The guy I disliked saw me, and it was instantly game on as he charged at me. I was warmed up and loose from running, so as he tried to tackle me around my midsection, I knuckle-jabbed him simultaneously with both fists in the carotid artery and jugular vein. We hit the pavement like a battleship anchor, and I sprang back up while he lay there, blinking and gasping. I snarled, "Get up! I'm gonna finish this!"

As he regained consciousness, he waved me off and said he was sorry about the game months before. I guessed correctly that he knew exactly what this was about. His friend stood there looking like a porcine gut with the shit slung out of it, obviously still half drunk from the night before. My friends snickered as they walked back toward the gate with the guy's buddy. I had no idea where the girls went and guessed they bailed as soon as the fight started.

I stood over the airman then, suddenly feeling like a real jerk. I held out my hand to help him up and asked if they wanted a ride back to the barracks. So I drove us to the chow hall and we five had breakfast together. It was a strange experience, and under different circumstances, we might have all been friends.

Of course, word got out because my day shift peers at the Main Gate talked about it. The story amplified that I took on four guys, fought like a wild animal, and then also banged their girlfriends — standard rumor mill nonsense. At work the next day, SMSgt Lippi had already heard and questioned me about what happened. I assured him it was nothing serious, just two GIs finally settling an old grievance and walking away on good terms. Lippi said, "I'm glad you prevailed, but you start out-processing in a few weeks. I don't want to see you get put on international or administrative hold, which could jeopardize your next assignment. If that happens, you could get redirected to a SAC base in North Dakota." I promised to behave and thought, if I get orders to a northern tier SAC base, I'll pay someone 1000 pesos to put me on international hold so I can stay here.

Around the time I returned to Alpha Flight, a Boston prep school graduate named Gary Bailey was assigned to Alpha Flight LE and lived across the hall from me in the barracks. He was a wealthy kid from Lexington, Massachusetts, who enlisted to avoid going to college. Gary wanted to get away and see the world. Another new guy named Matt Rossoni, a former Los Angeles cop, moved in next to him. The three of us became good friends and we enjoyed hanging out together.

Compared to the rest of us low-bred swine, Gary was tall, blond, and had a high-fashion taste, so he looked like he stepped off a recruiting poster. He wore Izod clothes and used grooming products I had never heard of. And women threw themselves at him. His grandparents sent him a monthly allowance (more than his military salary if I recall correctly), so he always had plenty of extra money. He let me borrow his "Preppie Handbook" by Lisa Birnbach; although I snickered about its fashion tips, I genuinely learned a great deal reading it. Gary had a video camera and shot a lot of tapes on our road trips. Yes, he may still have video evidence of me doing really stupid things in my wild, intemperate youth. Gary and I were on the same Alpha shift, so SMSgt Lippi directed me to train him and several other new guys. He was smart and caught on quickly, so he was easy to teach.

Sometime midway through my tour, I bought several grenades from the weapons dealer at the Nepo Market. Despite giving me a good deal for sending him so many referrals, I soon regretted buying the things. I wrapped them in plastic and buried them under the guava tree on the southeast side of Barracks 7504. Because I was leaving soon, I looked for an opportunity to use them. In early June, I was discussing local fishing spots with a PAF friend who was from the Mount Arayat area. He suggested it was easier to shoot fish while wading in shallow water than catch them with nets or tackle, implying his family owned private firearms. His grandparents owned a farm on the Pampanga River, and he invited me to go with him to "get some fish" one weekend, albeit it was clear I had to drive. The farm was surrounded by sugar cane fields, so we could either fish or shoot at fish without being bothered by anyone. It was about 35 miles away by road, so we agreed to go that Saturday since we were both off then.

Once we arrived at his grandparents' place, I surprised him by showing him the grenades stashed in my trunk. This made him happier than a bus load of sugared-up kids on their way to Chuckie Cheese. He knew a good deep hole with lots of fish where the concussions would stun them. We could harvest the big ones for his grandparents and leave the smaller fish. After all the hugs and excited conversing in Pampangan, he told his grandparents we were going fishing. His grandfather wanted to accompany us, and we brought him along. As his granddad began grabbing up gear, my friend stopped him to explain what we had in mind. The old fellow smiled and agreed on condition that I let him throw one of the grenades. That was fine with me since he was a WWII veteran who survived the Bataan Death March; I trusted his word that he knew how to handle "pineapples." We took rattan baskets with us and hiked out to the riverbank. Sure enough, there was no one else around, so we spread out and picked our target spots in a deep bend of the narrow river. On the count of three, we pulled the pins, tossed in our grenades, and ducked. They detonated almost in unison, nearly muffled by the 4-5 feet deep water, yet still strong enough to blow thick geysers into the air.

We rushed into the water and started collecting stunned bangus (perch) and tilapia. While none were giants, we collected enough big fish to feed two dozen people and left the smaller fish to recover. We gathered up other fish that appeared dead and then hauled them back to the farmhouses. Naturally, my friend's grandmother was very happy about all the fish and said they would dress out and sun-dry the majority. We set about cleaning fish and were done in an hour. She cooked some big tilapia for our lunch, so we enjoyed them and rice cakes washed down with San Miguels. The elderly couple invited me to stay overnight, but I politely declined since I needed to get back to work on Sunday. However, I promised to return another weekend. That made her and the old man smile.

We headed back to Clark, and I dropped off my friend at the PAF barracks. Already down to a few weeks before I transferred, we never got together again. Nevertheless, I was glad for the grenade fishing experience and pleased I could provide a Pinoy family with enough fish to feed them for several weeks.

I started collecting wine midway through my tour. Due to MCO regulations, I could buy two bottles of wine weekly (eight per month) on my "booze card." I kept my collection locked up in a wall locker, retaining all my receipts in case of a "show and tell" visit by Merchandise Control. Three months before I transferred back stateside, I suddenly realized I needed to drink one bottle of wine per day or throw a big party to consume them all. Granted, not all of my collection was "fine" wine, as many were bottles of Blue Nun white wine that I had bought on sale for $2 each. Matt, Gary, and I drank about 2/3 of my high-brow wine collection, and I gifted the rest to friends.

The other thing I realized was that I had accumulated a lot of t-shirts and other clothes. I had a pile of t-shirts from various club events and uniforms from all the sports teams I was on. I still had my initial issue of three sets of jungle fatigues, plus I was issued an additional set every six months. I also had two sets of tiger-stripe fatigues I had Dick the Tailor make me for hunting, one each for summer and fall. Desiring to keep the uniforms, I had to figure out what to keep and what to give away based on my household goods shipping weight limit. I decided to give most of my excess clothing to the houseboys. Concurrently, I knew they couldn't walk out the gate hauling bags of clothes. So when their shifts ended one day when I was off, I loaded the clothing into garbage bags and gave them rides to their neighborhoods. They were tickled to get the shirts and other clothes, and I was happy to donate them. I figured they could trade what they didn't need for items they did. I also gave a few shirts and hats to the kids at the back gates. Always look for the win-win.

In late June there was a rumor of a pending alert during our three-day break, but we were not put on standby. Gary and I had already planned a road trip to Subic Bay and Manila; Matt was on another shift and couldn't go, so we took Victor Williams with us. Victor was a black kid from Memphis who kept us laughing with his antics and jokes. We hit the highway to Subic Bay, stopping along the way for fresh fruit and to take photos. My friends (neither of whom had been in-country long) asked why so many locals laid out sheets or blankets next to the road to dry fruit. I explained the slipstream from passing vehicles helped keep flies and other insects at bay.

Gary was taping away with his video recorder, at times hanging out the window for better views. We went out to Grande Island and rented a beach cottage for one night. We had a blast swimming, snorkeling, and hiking around the island. I awoke before daylight the next morning to run on the beach. As I cooled off sitting on the beach watching the sun rise over Mariveles Mountain (Bataan Peninsula), I felt completely at peace.

The next day, we drove around the Bataan Peninsula to Manila. We arrived at dusk due to heavy traffic jams and got stuck in a bad one. Victor flagged down a Manila policeman and got him to help us through the Makati traffic in exchange for a bottle of Jack Daniel's we brought along. The cop sat on my hood, yelling and blowing his whistle at other drivers to let us through to a high-rise hotel we spotted. We checked in at the Manila Gardens Hotel and enjoyed the pool on an outside deck high over the city. The next morning, we had doughnuts and coffee at the Dunkin' Donuts next door and were surprised by how many Pinoy-Americans (mostly dual citizens) were in Makati. This was among the best road trips I ever had, although we almost ran out of gas en route back to the base; fortunately, we found a gas station in a small town south of Angeles City that was open on Sunday. The following week, Gary was reassigned to the Customs Section at the MAC Terminal and put on swings and mids, so that was our last road trip together.

One night, there was a high-speed chase off base that involved a motorcycle fatality. The local cops got after an American on a motorcycle. The biker cut through a housing area he obviously wasn't familiar with, probably thinking he could lose them. He sped down a long street and rounded a curve only to discover a T intersection too late. The biker lost control and his bike plowed into the concrete block fence on the other side of the street. Atop the wall fence was decorative rebar, and he was thrown head-first into it, cleaving his head almost to the jaw by the rebar due to his not wearing a helmet. The pursuing police were greeted by a grisly scene. I was told they had to cut the guy's head almost in two to extract the body from the fence. Play stupid games, win deadly prizes.

Shortly before I transferred stateside, a good friend and I decided to go out for a last beer together. We went to a "cop bar" and I left after a couple. He stayed on with some mutual friends. Not long after I departed, someone came in to tell him that a guy was outside tinkering with his motorcycle. He went outside to discover another American messing with his front brake. Without so much as addressing the guy, my friend kicked his ass right there in the street in front of everyone, then rode away. He never did find out who was behind it.

My final month at Clark AB flew by. I participated in one last "drag your bags" mobilization exercise two weeks before I started out-processing. That same week, a Traffic Management Office team packed up all but my essential items, and I was easily within my weight limit. I could have shipped more, but I stuck to my plan to give away everything I didn't need or want. I also had a two-duffel-bag limit for the flight home. I took several more small bags of clothes and items out to Negrito and Sapangbato Gates for the kids there. I had gotten to know some of these kids well and would miss their antics. While they were happy to get the gifts, a few cried as I drove away.

Numba One GI was going home.

198206: This is Senior Airman Ron Payne in front of Town Patrol HQ, taken in 1982 or 1983 after I rotated stateside. Ron left active duty to pursue a civilian law enforcement career on the West Coast. Although we picked on one another in the Philippines, we've stayed in touch and pester one another to this day.

I attended several ad hoc parties for outbound guys while out-processing and had mixed emotions about saying farewell. As had occurred with me 18 months before, I joined the NCOs who hauled several groups of FNGs downtown on their first nights in country. I gave them "The Talk" about social mores, manners, and things to avoid. As I said my goodbyes, I regretted not extending at Clark for another year. Everyone understood, of course. I needed to get back to The World for my own career development.

And then there was a dilemma about how to unload my M1911 .45 pistol. I wanted to sell it, but didn't want to risk returning to the Nepo Market weapons dealer or talking to an American who might rat me out. So I hit up my Constabulary friends about buying it. I only got 1500 pesos (roughly $150) for it and two full 8-round magazines, so I "lost" $100 for the 14 months I owned it. I considered this a positive investment return from a "safety and personal protection" perspective. That said, having fired half my original rounds during various road trips, the expended 16 rounds cost me roughly $6.25 each (factored for the $100 I went into the red). I marveled that I neither got caught with the unauthorized pistol nor used it in self-defense during my many forays off the installation. Unlike those who never left bars and shopping districts near the base, I constantly went places where Americans were scarce or didn't go at all. While traveling, that .45 was almost always within easy reach under a newspaper or my briefcase on the front passenger seat. In retrospect, I could have shipped it back to the US since the TMO guys packing my household goods shipment never checked anything I packed. Years later, I learned I could have removed the firing pin, declared the pistol a demilitarized historic artifact, and legally shipped it back to CONUS. Anyway, TMO came to pack everything two days after I sold the pistol. Double dookie!

As my tour wound down to its final days, I reflected on some of the guys with whom I served. While some were memorable for personality quirks, foolish choices, and unfortunate incidents, I remember others due to photo captions or notes I jotted down and discovered years later. A few noteworthy people follow, most of whose names and identifying information have been changed to protect the guilty.

Michael "Mikey" Mulberry (not his real name) was from a small town in the Ozark Mountains. He was affectionately and otherwise known as "Mike-Hell" — a nickname well earned by his escapades. He possessed above-average intelligence and was OK at his job, yet was a loose cannon when fortified by liquid courage courtesy of his buddy Jack Daniel of Lynchburg, Tennessee. Even when sober, Mikey had an unpredictable streak, as evidenced by his refusal to call for fire assistance when a black market facility caught on fire outside Dau Gate one night. He unplugged the phone so the PAF guards couldn't call for help. He was never disciplined for that stunt since it wasn't a USAF issue and burned up a lot of contraband items. Mikey had a penchant for — as one NCO eloquently described — "letting his twenty-dollar ego and mouth outrun his ten-cent ass." He cussed out a higher-up or three and God knows how many other people at the NCO Club during Country-Western nights, most of which got him ganged up on and monkey-stomped. Another time, he "accidentally" spat tobacco on a dependent wife during a confrontation at Clarkview Gate; however, that neither excused him for chewing tobacco on duty nor washed the Red Man juice out of her dress. He received a "suspended" Article 15 (nonjudicial administrative punishment) for that one, which would have gone away with 12 months of good behavior.

Perhaps Mikey's great claim to fame was receiving — not one, but two Article 15s in a single 12-hour period. The first was for swinging on the wide gate at Mabalacat Gate during a swing shift. The ground dropped off steeply into a washout just past the pavement, so when the gate was swung wide open, it was actually about 8 feet over the sloping ground at the far end. Bored guards would sometimes swing on the gate, but when caught, it was an automatic Article 15 penalty. As predicted, he was caught swinging on the gate around 2100 one night and got chewed out royally by his chain of command.

Instead of going to his barracks room after the shift and sleeping off his perceived persecution, Mikey went to the Peacekeeper's Pub with a bottle of whiskey and started drinking heavily. During the wee hours, he began flipping over furniture and throwing things, so the snack bar attendant called the LE desk. By the time a patrol responded, he had picked a fight with the much bigger CG, who pounded him cuckoo for Cocoa Puffs and handcuffed him. Mikey received a second separate Article 15 for this episode of drunken foolishness. I seem to recall that the second Article 15 was already processed before the first one made it to Major Rich's desk.

Another time at the NCO Club, I spotted Mikey across the ballroom dressed in his finest cowboy clothes and an expensive black

Stetson. He was on the prowl for some lucky urban cowgirl and would make — in his Ozark colloquial — "a sho' nuff best effert" should one return his longing gaze. He leaned back so far in his chair that I expected he might fall over backwards at any time. Then a moment of pure magic occurred when a woman's eyes met his. The woman wasn't bad-looking until she flashed a grin straight from a periodontal nightmare. A tablemate observed, "She's a human jack-o-lantern." We saw Mikey grimace and attempt a hasty exfiltration as the woman swooped in, Marlboro first like an F-4 Phantom on a low-level strafing run. Mikey started to toss down his drink as he jumped up, only to hit the brim of his Stetson and splash his own face and the front of his chamois shirt. The woman stopped in her tracks, then laughed and scurried away to flee his scowls and curses.

Another time, Mikey was his usual drunken self, walking up through the NCO Club ballroom's seating levels, when someone yelled out to him from a distance away. As he turned to raise a glass to the other fellow, he tripped and fell "up" the steps and into a table. Of course, everyone who saw this laughed, so he turned and bowed to the crowd before continuing on. I could tell so many funny yet sad stories about this guy. He got into so much trouble in so little time. Several others and I tried to mentor Mikey, but he was extraordinarily stubborn, and the lessons just didn't take. The chain of command finally had enough and dishonorably discharged him following his third Article 15. Poor Mikey didn't last even a year in-country.

Thereafter, his memory spawned warning vignettes to FNGs about staying wired tight and avoiding foolish decisions. Mikey Mulberry remains the lone individual I knew during my three-decade career who managed to earn two Article 15s in a single 12-hour period.

Dean Bidwell was my barracks roommate for several months, then my "ghost" roommate for half my tour after he moved off base. Although on different schedules, we hung out some off-duty when our off-duty days coincided. Dean was intense at times, yet loyal to his friends. He was from Kendallville, Indiana, and definitely the "Northern" type, so the tropical climate was an adjustment for him.

Dean wanted to join Horse Patrol — the "Clark County Mounties" — and his first two requests were disapproved. He took riding lessons at the base stables, studied equine anatomy and literature, and took several tests. Finally, he was granted an interview, and his transfer request was approved. All went well for a time, and he loved the duty. Then bad things started happening to him.

One midnight shift while eating chow at Sapangbato Gate, a power blackout occurred, and the generator next door, FEN-Radio, screeched as it kicked on. The sound spooked his mount, and as Dean tried to control it, the horse slammed its head into Dean's face and broke his nose. He held the reins as the horse started to run, so he wisely let go. I don't recall where they caught the horse, but Dean was injured and shaken up.

Another time, he was patrolling close to the runway's north end near where SrA Sam Gray was killed in the late 1970s. Something spooked the horse, and it took off wildly toward the washouts. Besides the steep drop-offs, some washouts still contained old concertina and barbed wire from WWII. Dean tried to guide the mount clear of a large washout, but into it they plunged, and both were injured. He was hospitalized for several days and relieved of duty from Horse Patrol. I regretted it so much for him, yet felt he was better off working regular LE duty where he was less likely to get harmed. After the second injury in a short time, I sensed he probably agreed.

Dean and I were on the 13th Air Force Honor Guard together and performed at many ceremonies and parades. Because his girlfriend (later wife) was Pinay, he learned a lot of Tagalog quickly and taught me some. We liked to sing the local advertisement jingles, including that of San Miguel Beer — "Mag beer muna-tayo!" We also mimicked the San Miguel Gin slogan: "Ginebra San Miguel — Wala-ng katapat!" That meant "San Miguel Gin — There is no other!" When introducing myself to Pinoy strangers, sometimes I added "Wala-ng katapat," and that made most Filipinos howl with laughter. Even when I butchered their native tongue, most found it highly amusing that an American would try to incorporate Tagalog marketing slogans into casual conversations. I heard Pinoy say, "Komedyante ka" (You're a comedian) to me many times. Dean was always up for making our Filipino friends and PAF colleagues laugh.

Sam Smith was in my SP Academy class and ended up on A Flight LE after a 90-day temporary stint with the 7th SPS to work the flightline. His father was from Alabama and married a Briton while stationed in England; he either got out or retired in England because Sam was raised in the London area. I recall thinking it was very cool that this tall black kid had a British accent. He also spoke elementary French, so occasionally we spoke it to keep our conversations private. Sam called people "wankers" and "buggers" a lot, so these Limey slang words worked their way into my vernacular. He married a Filipina, and they had a couple of children before they parted ways. Sam left active duty a few years later and eventually retired from a small town police department in Pennsylvania. He can always count on me and his buddy Mark Hawkins for snide comments on his social media pages.

198203: A1C Samuel Smith in March 1982. Son of a military retiree who retired in England, Sam and I could speak French to one another and confuse eavesdroppers. I learned many British slang terms from him.

Mark "The Hawk" Hawkins arrived in the Philippines a few weeks after I arrived and drew the short straw by moving in with Mikey Mulberry. He soon moved in with Sam Smith, and they became close friends. Hawk had — still has — a larger-than-life personality, but at the time he was 6'2" and very thin. He was smart, dependable on the job, and resourceful. He somehow managed to buy a jeepney and registered it on base. Hawk also married a local and had a couple of kids before they also parted ways. He lives in the Atlanta area and travels a lot, so we occasionally get together whenever possible to eat seafood.

One rainy afternoon, a muscular, articulate black guy moved into the barracks room next to mine. Will Burgh was in his 20s and seemed to have a peculiar aura, as if he were on a mission instead of someone who had just completed LE school. He obviously sensed my suspicions, and we talked often, especially about college sports. He and his roommate, Peterson, played basketball most afternoons on the court in front of our barracks. Occasionally, Peterson acted as if he knew something about Will that I didn't, but both were solid guys, so I didn't inquire. Burgh grumbled about things and acted impatient at times, like he was waiting for something. Within a couple of months, both Burgh and Peterson were working in Customs. I figured both were smart guys, and that frequently led to special duties and administrative jobs.

After that, I didn't see much of either, and I soon moved to a different room downstairs. Months later, I was back stateside, and a friend wrote to me about a big bust that went down in Customs — drugs, if I remember correctly. Concurrently, a big theft ring was interdicted in the Supply compound. It turned out that Burgh and possibly also Peterson were undercover agents from the Air Force Office of Special Investigations. The missing pieces to their purpose mosaic fell into place. One of them later won a national law enforcement award.

Mark Waters (Hurst, Texas) was a year ahead of me and the son of a retired CMSgt. While he was sometimes abrasive with people who crossed him, he was a decent LE troop, feared neither rank nor a person's size, and did his job well. I'm not sure where he lived on base because I rarely ever saw him off duty. He had previously worked with Don Quesnell on youth outreach programs. Mark didn't get along with my roommate Dean, and they almost got into a fight in the chow line one day. He served on A Flight LE with Don Adams and me; we always got along well. Coincidentally, we were all three sent to Carswell AFB for our next assignment, so Mark and Don were 30 minutes from where they grew up. Somehow, we were all three assigned to D Flight LE there and later became roommates.

Don Adams was from Irving, Texas (as he loved to say, "Where the Dallas Cowboys play"), and his father was a Korean War veteran. He was a big fellow (6'4" / 225), a great LE patrolman, and among our flag football linemen. While I could run circles around him, if he ever caught me, I was in trouble because he was very strong. He liked his tequila straight, made fun of Yankees, told jokes in Spanish, danced the Two Step at the NCO Club's country-western nights, and brawled with people he didn't like. I got him to join the 13th Air Force Honor Guard, and he served on it for a while before transferring stateside. Don's best friends were all Texans — Greg Haselberger (Houston),

Dwayne Lowery, and John Marino (Dallas). I'll share more on Don in my Carswell AFB chapter.

While I knew a lot of the Pinoy in our DOD police force, my favorite was a guy named Exiomo. He was a creative fellow, a terrific artist, and a jack-of-all-trades who won a base-wide crime prevention poster contest. That got him noticed, and thereafter he did a lot of graphics work for the 3rd SPG and 3rd CSG. I liked Exiomo because he was a pleasant guy and joked around a lot. Like most other Pinoy working on base, he called me and everyone else "Sarge" regardless of what rank we actually wore.

198202: Department of Defense employee Mr. Exiomo and me standing outside the Pass and ID building in February 1982. Exiomo was a fantastic artist and did a lot of work with us.

Steve Borden (Tampa, Florida) was my SP Academy roommate and roomed with Bruce Bielby (Bethany, Missouri) since they were on Charlie flight LE. None of us requested overseas duty — much less the Philippines — yet we were all glad for the experience. Our shared personality quirk was our dark sense of humor, and between us, someone always had a terrible joke or five. We delighted in calling one another a long list of nasty names for fun. Steve and Bruce used to joke that they were a couple to scare off the bar girls when they went into town. Both had significant others back home whom they intended to marry, so they only went downtown to drink or watch the nightly freak shows.

One morning, I saw them drive up to the BCA to turn over their jeep to our shift. While the jeep was somewhat clean, they were splattered with mud. They were allegedly chasing some intruders through a field and nearly got the jeep stuck, so it was covered in mud before they extracted it. Their flight chief was very upset and made them clean the jeep, but refused to let them change uniforms at the barracks — didn't bother them at all. A few months later, Steve transferred to MacDill AFB in Tampa and served out his remaining enlistment there. Bruce went to Wightman AFB, Missouri. After he got out, Steve married his girlfriend and served as a Tampa Police officer for several decades. I ate lunch with him a few times while stationed in that area 20 years later. Bruce attended college, got married, and was a high school teacher for many years. We all stayed in touch and still swap barbs unmercifully with one another.

At age 25, Oscar Reid was an "old" enlistee and already married to his wife, Melanie. He grew up in Abbeville, South Carolina, and his father owned a car dealership. He was on A Flight LE and mature beyond his years; he always appeared immaculate and was very smart with a diverse taste in humor. He volunteered to work with our Target Hardening initiative in the Lily Hill housing area, so we made a solid team. What he learned with Don and me was far more useful vis-à-vis career development than waving cars at a gate. Oscar was like a sponge and soaked up information as fast as we could teach him. He went on to work Horse Patrol for most of his three-year tour there. We stayed in touch, and his next assignment was to Gila Bend AFB, Arizona. I sent him a few funny cartoons warning him not to get bitten on his private parts by rattlesnakes. Oscar and Melanie settled in Due West, South Carolina, and bought a classic three-story antebellum home next door to Erskine College. Oscar ran a radio station for a while. They returned to the Philippines for missionary work at least once and had recently arrived back home from a trip when Mount Pinatubo erupted in 1991. I stopped by and visited them in the late 1990s while visiting my Clark ancestors' graves in that area. We're still in touch, and I hope to see them again.

One Saturday night, a couple of weeks before I transferred back stateside, I got this crazy idea that I needed to cut a donut on the parade ground. With the Officer's Club and the single officers' barracks (aka "The Zoo") adjacent to it, there were always stories about inebriated junior officers occasionally pulling such stunts. They typically got caught, either in the act or more often via post-stunt boasting. With the parade ground far from any enlisted barracks, no one would suspect a junior enlisted guy. I knew where the barracks CQ stashed the keys to one of the command's staff cars. The impulse to take it was irresistible. There was also an extra radio I could use to monitor any responding patrols. Guys assigned to CQ duty were typically the "sick, lame, and lazy" (SLL) or others on limited duty due to injuries or accidents. Despite the requirement to make hourly rounds of the entire barracks complex, the SLLs rarely did, and most just "pencil-whipped" their checklists. There were always people coming and going at all hours anyway. This particular weekend, the CQ was a slovenly, shiftless turdling who stayed sacked out in the Peacekeepers Pub's TV area. He un-parked his lazy ass only to use the latrine or when overwhelming hunger stabbing his guts compelled him to go eat. Discovering he left the door to the CQ office unlocked, I used his complacency to cover my tracks. Knowing the midnight chow schedule and guessing the CQ would walk to the chow hall to eat, I discreetly grabbed the keys and radio as midnight chow got underway. As expected, he finally got up and plodded toward the chow hall three blocks away. It was five minutes after midnight. An off-duty Coonass friend named Boudreaux spotted me getting into the staff car and demanded to know where I was going. I told him I was asked to run an errand. Obviously seeing through it and sensing a memorable prank, he said, "You don't lie well, Clark. Either I go with you, or I rat you out." He was serious, so I relented, and off we went. We drove past the chow hall and counted the LE patrol cars. As expected, about half the shift's cars were there, along with both Enforcer patrol jeeps. We continued on toward the parade ground, and just before arriving, that sector's patrol car passed us heading toward the chow hall. The guys inside never looked at the staff car as we passed. I made one full circle of the parade ground and saw no other patrols anywhere near it. The whole area was deserted except for a few cars in the Officers Club parking lot and staff cars parked in front of 13th Air Force HQ. I turned onto the road intersecting the middle of the parade ground. Driving slowly to the middle of the rectangle, I whipped onto the grass and gunned the engine. Dust and grass flew wildly like rooster tails behind the staff car. I spun the steering wheel and started cutting the donut, just like Bo and Luke Duke in the Dukes of Hazzard TV show. As Boudreaux and I laughed hysterically, I reversed direction and cut another big donut. With two cut and knowing a patrol would get dispatched any minute, I turned and quickly exited the parade ground. Almost simultaneously, the radio crackled to life. The LE desk sergeant called for any available patrol to intercept a car cutting donuts on the parade ground. Three patrols answered, and one reported that it was en route from

Sapangbato Gate only six blocks away. Another patrol was inbound along the route back to the barracks, so I couldn't get out that way. Needing a place to hide, I turned down a side street, pulled into the back exit of the base commander's HQ, cut the lights, and parked next to the base commander's staff car. Within a minute, three patrol cars were circling the parade ground. Dust still hung in the air, and two patrols peeled off in the direction of the most likely escape routes—both toward officer housing areas. We sat there for a few minutes watching the remaining patrol car. Whoever called in the incident either didn't give a good description or didn't see where we went. Obviously, the responding patrols never thought to look next door and check the base commander's HQ parking lot. Although we were parked in plain sight, the patrols drove past us without looking our way. The lone patrol soon called the desk, reported that the perpetrator was long gone, and also drove away. We sat there laughing as I waited a few more minutes to drive back to the barracks. I then took an indirect route back past the Airman Club, 3rd SPG Training building, and the Supply complex. This way I passed behind the Base Engineer complex and only had to cross the main road right in front of the Peacekeepers Pub. When we arrived back, I told Boudreaux, "You keep your mouth shut about this until I'm out of the country. If I go down, you go down too." He nodded, and I knew he'd keep quiet. I wiped my prints from the steering wheel and gear shift, then returned the radio and keys before walking into our barracks pub. As I entered, I saw the CQ sitting on his tail watching TV and said hello to him. He mumbled something unintelligible and yawned. When I asked if he was having a boring night, he nodded without taking his eyes off the boob tube. That dumbass never even heard the radio report or knew the staff car was missing, I thought. I grinned as I left the pub for my barracks room, pleased to have pulled off yet another memorable stealth mission.

I took one last tour around the base the week before I left. I snapped three rolls of photos of various places and people and stored them in my camera strap for processing. The conflicting emotions were intense since I had — in most ways — matured so much during my tour there.

As I closed out my postal box, the clerk handed me my final mail, including a letter from Hannah Hill. She said she would serve as a counselor at the 4-H camp in Columbia, TN, the week I arrived home and wanted me to visit her there. I didn't have time to write back since I was flying home in a few days, so I decided I would just show up there and see her. The day before flying home, Matt Rossoni and I drove to Merchandise Control, where I signed over my car to him to sell for me. It needed a few minor repairs, and he would extract the cost before mailing me the check when it sold. It was hard to leave my friends, but at the same time, I was eager to get home again. Saying goodbye to Don and Belle Quesnell was especially hard. But I had to move on now, and getting home safely dominated my thoughts.

I awoke before daylight on July 27, 1982, my final morning at Clark AB. I went for a two-mile run around Barracks Row in a heavy mist. It was hard to fathom how quickly the 18 months had passed. My Alpha LE buddies were driving in for guardmount and honked their horns or yelled stuff like "See you in hell, asshole" and "Goodbye, dickhead" at me as they passed. Not wanting to take wet things on the plane, I gave my running clothes and shoes to my houseboy, and he seemed happy to get them. I figured after washing, he would sell or trade these items. He and the other houseboys acted sad to see me go. They asked me to stop by and see them when I got stationed in the Philippines again. So many people believed I would return.

After breakfast, Matt drove me to the MAC Terminal and stayed until my flight boarded. The Flying Tigers 747 had arrived the day before and sat mutely on the ramp, enveloped in light mist. John Marino came by and brought a final round of San Miguel beer. Gary Bailey was working the terminal desk and checked in my bags for the drug dogs to sniff. Other friends stopped by, including a few patrolmen from Alpha. I knew Don Quesnell was patrolling on the other side of the base and wouldn't be able to make it. My boarding call came, and I waved goodbye as I walked out the terminal door to the waiting plane. I turned at the top of the stairs before getting into the aircraft and saw Gary Bailey salute me through the big plate-glass window. I returned his salute and stepped in.

I was seated in the rear of the plane on the left side. I didn't know anyone around me, so I had no one to talk to. The seats next to me were empty, so I moved to the window. Fifteen minutes later, the engines fired up and we taxied out. Even though I knew the flightline well, the planes and buildings looked strange as they passed, as if I had never seen them before. By now it was raining outside and I couldn't see much. The big jet turned at the hammerhead, roared into the air, and all I could see were heavy clouds and threads of water zipping across the porthole. We climbed hard and soon broke out of the clouds into brilliant sunlight, the horizon split into halves—a fluffy white bottom topped by a blue canopy.

A few hours later, we stopped in Okinawa to pick up more passengers, mostly Marines. It was pouring rain there too. As we lifted off, the Jarheads cheered—obviously glad to leave Japan. I spent the next few hours reading magazines and watching an in-flight movie but could not sleep. What I did enjoy was watching the four-hour sunset between Japan and Alaska. Our flight path was far enough to the north that the twilight never completely ended.

As our plane flew east, we "lost" hours and then a calendar day by crossing the International Dateline somewhere over the Pacific Ocean. We landed in Anchorage, and I said hello to the big growling polar bear from 18 months earlier. I grabbed some chow at the snack bar and saw a couple of SPs I knew who had boarded in Okinawa. They were heading to SAC bases in places with cold climates and thought I was privileged to get one in Texas. While they complained about their assignments, another airman overheard us and said, "You're all so lucky. I requested a consecutive overseas tour to Europe and got Thule, Greenland." Looking at his disgusted expression, I felt sorry for him and was grateful for my assignment. Our next stop was LAX in Los Angeles. We were there for a short time, and I called my parents collect to confirm my landing time in Nashville. We next flew to Lambert Field in St. Louis, where we reclaimed our baggage. I quickly discovered a problem I hadn't thought through: how to carry two duffel bags and my carry-on satchel across the terminal. A sailor heading my way offered to carry one of my duffel bags, so I gave him five bucks to buy lunch. I made it to my connecting flight just in time. About ninety minutes later, I landed in Nashville, and my mom was there to pick me up. I was glad to see her, despite her immediately remarking that I smelled bad and had a strange accent. I asked to drive and almost got into a wreck within five minutes, so I pulled over and let her drive. I had left the Philippines almost 24 hours before, and due to adrenaline and excitement, I had not slept except for short catnaps on the flight over the Pacific Ocean.

Around 2200 hours, we arrived home to the farm. Our big Collies met me at the car and cautiously sniffed me over for a moment before jumping up on me. My dad and brother were home and gave me big hugs. Shaen had grown a lot while I was away and was almost my height now. Exhausted, sore from sitting, and still acclimated to a time zone on the other side of the world, I didn't socialize long. I grabbed a few bites to eat and headed upstairs. After taking a shower, I crashed out on my bed and slept for over 16 hours. I woke up at 1600 the next day to the sound of cows mooing and mockingbirds singing in the maple trees outside my window. It was good to be home at last.

"Zoomie Pig Odyssey, 1980-84"
Carswell AFB, 1982-84

While I was glad to be home from the Philippines, I quickly discovered I was out of place. I had changed in ways beyond my own comprehension at the time. Whereas the lives and lifestyles of people at home were mostly the same, my relations to them — how they knew me — had frozen in time when I left two years before and were far different when I returned. I was older, street-smart from police work, and had grown physically — bigger, stronger, and faster. In addition to adopting a military culture and lifestyle, I had lived in a foreign country and absorbed some aspects of its diverse culture. According to family and friends, I spoke with a strange accent and was odd. I had been exposed to and had participated in violence, and in a disturbed way even enjoyed the thrills and adventure from it that no athletic sport could provide. I was glad to see my family but quickly grew bored being there. To them, I was "still just a kid," but that reality was long past.

Beyond that, I discovered that being around my high school friends was disenchanting because most had changed so little. It intensely annoyed me how so many would ask me about my experiences and lose interest after one minute. They wanted sound bites, Cliff Notes, and quick summaries of complicated events and situations that were difficult to explain. In retrospect, they lacked a cognitive framework to understand any of it. To everyone except those who were veterans, I was still the same person they knew from two years before. Yet I wasn't.

Not even close.

Two days after arriving home, I drove up to Columbia to see Hannah Hill at the 4-H Camp. I stopped to check in with the camp manager, Boone Morrow, who was glad to see me and said the campers were at the pool. I walked out to it and saw Hannah directing kids in a swim competition. She didn't see me at first, but when she did, she froze and just stared. We ate dinner in the mess hall and went for what turned into a long walk. We talked mostly about old times, her college classes, and so forth, and a little about where I had been. Already old friends, I felt at ease with her, and she confided that I was very special to her—that she wanted us to pursue a relationship. I agreed, and we planned for me to come visit her at Western Kentucky University that fall when I could schedule leave from home. Around dark, we said goodbye, promising to write, call, and stay in touch. I drove home completely at peace.

I could have taken 30 days of leave but decided to save it for later. I cut it short at 10 days, loaded my car, and headed for Texas. Departing at daybreak, I drove cross-country and arrived at Carswell

AFB around 1600. I signed in with my new unit, the 7th Security Police Squadron, to stop my leave. I walked over to the SP barracks, got assigned to a barracks room, and unpacked. My roommate, Seagrove, lived off base at his own expense, so I had the room to myself for the seven weeks I lived there. Scheduled to pin on E-4 (Senior Airman/SrA) on 1 October, I would qualify for housing allowance and could move to an apartment off base. Carswell was a B-52 bomber base transitioning from Vietnam-era B-52Ds that delivered traditional gravity bombs (conventional and nuclear) to B-52Hs armed with Air Launched Cruise Missiles (ALCM). The 7th Bombardment Wing was part of Eighth Air Force under Strategic Air Command, respectively headquartered at Barksdale and Offutt AFBs. The B-52 alert crews had the run of the base and could park anywhere they pleased. Outside most facilities was a long red-painted curb designated as alert crew parking only. Anyone parking there other than an alert crew received a DD 1408 traffic citation on the spot.

I spent the next two weeks with in-processing and refresher training. Assigned to Delta Flight, I worked a split shift: three swings, three mids, and three days off with a "24-hour break" between the swings and mids. My flight commander was Captain Donald Neal, an Air Force Academy alumnus who graduated at the top of his class. Neal qualified for any branch he wanted and chose Security Police; he intended to get out after his five-year statutory service obligation and pursue a law career. He was the most professional officer with whom

I served during my enlisted years. My flight chief was TSgt Robert White, an older NCO nearing retirement. He was a Vietnam veteran who had also served in the Philippines. TSgt White and I got along well, and he felt my Philippines experience benefited the team.

Delta Flight had 70-80 SPs assigned to it, split between 55-60 Security, 15-18 Law Enforcement, and 2-3 K9s. Security was responsible for everything inside the flightline, and LE had the rest of the base and cantonment area. The entire northern perimeter of the base was Lake Worth, a dammed section of the Trinity River's west fork. Across the lake was a high sandstone ridge covered with high-priced homes, oaks, and mesquite trees.

From west to east along the lake, responsibility for the airfield's north hammerhead and Weapons Storage Area (WSA, the nuke dump) belonged to Security; the marina, picnic grounds, and Officers Club area belonged to LE. Everything else along the perimeter until it reached the flightline again on the base's southwest side also belonged to LE. The base hospital on the northeast side actually fell outside the base during its massive renovation. After completion, the hospital grounds were reincorporated inside the perimeter fence and the gate shack moved several hundred yards to the southeast.

South of the hospital, base housing was inside the perimeter fence down to the Trinity River. Turning southwest, the fence followed the river to the NCO Club and then southward to Highway 380. Aside from the East Gate (manned 0600–1800 M–F), two barracks, a supply parking area, and an indoor firing range, nothing else occupied the eastern perimeter south of the NCO Club until the fence met Rogner Drive near the Main Gate. The off-base NCO housing area was on the south side of Highway 380, and LE had concurrent jurisdiction for it along with the Westworth Village Police. Many on its force were former Carswell SPs or AF Reserve SPs.

The Main Gate was the focal point of the south perimeter. Next to it was the Pass & ID building, a picnic area, and the base commander's house. From there, the fence ran west along the officer housing area to the West Gate a half-mile away. It was manned 0600–1800 daily to accommodate those in the housing areas; with no shade, this gate was blistering hot on summer afternoons. From there, the perimeter swung around to the southwest and south past more base housing until it reached Highway 380. From there, it ran west to the runway's south hammerhead. Security resumed responsibility at that point and back northward to the lake. The General Dynamics aircraft plant occupied the entire western side of our perimeter. It had its own high-security operation to protect its property and assets.

Outside the southern cantonment area (main base) fence were more NCO housing areas, the golf course, more officer housing, and the base elementary school. Although these areas were outside the cantonment area fence, they fell inside another outer perimeter fence.

One had to access this area either from two roads outside the Main Gate or at a specific stoplight on Highway 380. Base officials decided to stop manning several gates in those areas a few years before. However, the areas were clearly marked as base property, and anyone entering was subject to search. The signs and our visible patrols probably discouraged most without business from driving in. While the K9 kennels were also outside the perimeter fence on the golf course, our K9s weren't exactly a theft hazard. If someone tried to steal them, I'd have gladly paid to watch that drama. Usually, a K9-qualified senior NCO served as the kennel master, and this NCO had a kennel attendant on every shift. Often the attendant was an SP who was on light duty due to convalescence or someone assigned temporarily after getting in trouble.

Something I anticipated was a former colleague trying to interfere with me from the Philippines. He allegedly contacted his old friends in the Training Section to tell them I was untrustworthy. Several NCOs evidently believed him, since they treated me as if I were a pariah. I don't know what all he told them, but I heard none of it was positive. I soon found myself in front of LtCol James Griffin, the 7th SPS commander. Without divulging anything he had heard, he warned me that I was under close scrutiny and should stay out of trouble. The lone training NCO with anything positive to say was MSgt Reggie Phillips, who looked over my records and liked what he saw. He believed nothing he had heard; he recommended I disregard what anyone else thought and set about proving myself. TSgt White said the same, so I had at least two influential NCOs in my corner. Captain Neal voiced the same assessment, and I sensed he would give me a fair shot.

Two other SPs in-processing with me had reported from overseas tours. I quickly bonded with Ronald Rucker, who had served at RAF Greenham Commons, England. Married with a two-year-old son, his family would join him in a couple of months. I didn't enjoy hanging around at the barracks, so I visited Ron at his apartment a lot until they arrived. Ron and I were both headed for Delta Flight. I don't remember the third guy's name but recall he served in Germany.

The Training Section wanted to use us three overseas arrivals as opposing force (OPFOR) cannon fodder for the Peacekeeper Challenge (PC) team, currently training for the upcoming Air Force competition at Kirtland AFB, NM. We were to engage the PC team with MILES (laser) gear at the Non-Nuclear Munitions Storage Area (NMSA) 10 miles west of Carswell, out past the General Dynamics F-16 plant across our runway. Our four-man PC team (all young NCOs) bragged about how great they were, predicted we were no match for them, and guaranteed they would kill us off quickly in every scenario. Ron and I had other plans.

I had used MILES gear several times and liked it. Each time you fired a blank round, a barrel-mounted box emitted a laser pulse. Each man wore a helmet and vest with sensors that would detect a near-hit or "kill" shot and transmit these to a small speaker on the vest. Near-hits sounded as quick beeps, whereas a kill was a steady, shrill "beeeeeeeeep" that was very loud and annoying. You turned it off by removing the firing key from your emitter box.

When I showed up in tiger-striped fatigues and face paint, the PC team laughed at me. "Hey, Rambo, who you gonna kill?" "Where'd you get those pretty pajamas?" "Have you ever even fired an M16 before?" It went on and on. When Ron and I attempted to dish it back to them, the dickheads tried to pull rank as if that was necessary. The PC coach, an older NCO and Vietnam veteran, was the only one among them not laughing. He looked at us and said, "Y'all know what to do." I then realized he picked us for a reason.

We got to the NMSA around 0800 and it was already hotter and more humid than a blown radiator on a backcountry dirt road. We were to run three scenarios. The first was the PC team reacting to intruders inside the wire. We would then trade roles for the second and attack the defending PC team. The final scenario was a repeat of the first. The Training NCO drove the "Peacekeeper" (armored flightline truck) and graded the scenarios.

We had five minutes to plan for the first scenario. We decided to split our three-man team into individuals. We would hide and let the PC team drive around a couple of times until they exited their Peacekeeper to deploy. Then we would hit them in the open. Ron hid atop the corner (first) igloo, while I hid inside the drain culvert under the east internal road to await our third guy creating a diversion. Number three (our diversion man) would lie down in a depression by the perimeter fence, then fire a few shots to provoke the PC team into deploying on him before surrendering. When that occurred, I would exit the drain culvert and sprint around past Ron to the second igloo, where I could then cover his rear. Ron would fire several shots to force the PC team into flanking him. We figured the PC team would assume both Ron and I were atop the south end of the first igloo. Once they committed, I would engage them from behind and bag all those on my side.

Our plan worked perfectly. Spotting our diversion man, the PC team drove straight past my culvert without checking it. As they deployed on him, Ron popped up and shot one in the back from atop the igloo. As the others took cover and shoved our captured diversion man into the Peacekeeper, I sprinted across the road past Ron's igloo to the second one. Since I ran past in their blind spot, they never saw me and fired blindly at Ron's igloo. They think we're both on top of it, I mused. I could hear Ron laughing. "Hey Dan, I got one — three to go!"

I moved to the front of the second igloo and climbed atop it. One PC team member kept Ron pinned down on the east side as the other two maneuvered around the west side of his igloo directly across from me. I gently rose to eye level behind the grass and saw Ron engaging the single PC guy, fixing him in place near the igloo's front corner. The other two were 40 meters away in the open, so I flipped my selector switch to full auto and opened fire. The steady beeeeeeeep sounds indicated unmistakable kills. Realizing he was now alone and trapped in the open, the last PC member tried to escape to the Peacekeeper vehicle, but Ron and I bagged him in a crossfire. The Peacekeeper sat idle for a few minutes and then drove away. The PC coach called ENDEX and frowned at his team, declaring us the winners of the first round. As I strolled past the four "dead" NCOs and their angry glares, I laughed and said, "Hey Ron, check out all these dead rookies!" Their cursing-filled replies are best left to the imagination.

For the second scenario, the Training NCO loaded us "insurgents" into the Peacekeeper and drove us back to the guard house. The NMSA guards heard everything over the radio and were cheering for us. Giving us thumbs up, they were happy to root for the underdogs. I asked the Training NCO what we were supposed to do. He replied, "Avoid their mistakes when you deploy on them."

The radio soon called for us to respond. The Training NCO said we were in charge, so we asked him to drive around the igloos until we spotted them. We soon did but couldn't figure out how to deploy from the vehicle without getting shot. I'd never been inside a Peacekeeper, and it had huge blind spots due to the very small bulletproof viewing ports. The NCO refused to drive off the road between the bunkers, and that limited our options. We made a second pass, and the PC members we could see had split between the tops of two bunkers. Ron suggested we open a firing port to shoot at them when they popped up to fire. No sooner had I gotten a port open than a rifle muzzle was shoved in and fired off a 30-round clip. One PC guy had rushed up as we passed and was jogging behind the vehicle waiting for such an opportunity. While we lost Round 2, we learned a valuable lesson. Naturally, the PC team was howling, slapping high fives, and crowing about our loss after their embarrassment in the first scenario.

We traded places again. This time we split between opposing igloos, each second from the end of their respective lines of igloos. I set up on the one nearest the north perimeter road; Ron and the other guy were on an igloo in the row across from me. They purposely let the team see them, and I let the Peacekeeper pass by, listening for opening doors. The PC team dismounted from behind the first igloo, deploying to the second in two-man teams using a bounding overwatch. They sprinted across the open space as the overwatch team fired on Ron. Hidden from the overwatch team's sight behind the curve of my igloo, I bagged the first team crossing the open space with several aimed shots.

Our third man exposed himself, and the overwatch team picked him off. The team remounted and redeployed in front of Ron's bunker. I yelled for Ron to run to me, but he stayed put. Now under fire from behind the blast curtains on each side of his igloo, I low-crawled up behind the rear vent of my igloo for cover. Ron was cornered, and I couldn't move without hearing steady pops and "zips" from near-hits. He and a PC member traded shots and took each other out. It was now a one-on-one fight. After a few seconds of silence I glanced around the vent and could see the Peacekeeper driving away. I heard a voice yell "ENDEX" and saw everyone across from me walking to the road. The third scenario ended in a draw.

The after-action review (AAR) was interesting, to say the least. The Training NCO went over the lessons learned and thanked us for participating. The PC team rode back in the Peacekeeper, and the Training NCO took us back in a van. After we returned to the 7th SPS HQ, he pulled me aside and said, "Y'all went 1-1-1 against an experienced team. If you had not just arrived here, I would put you on the team right now. I hope you'll try out next year." The PC team members were friendly to us after that. After the Training NCO reported our OPFOR performance, the Training Section staff treated us differently too. MSgt Phillips said he wanted to get me on Delta's EST team as soon as possible. Every shift had its own EST squad, so I was happy to join up.

With training completed, Ron and I reported to Delta Flight during the first mid-shift of a cycle. We were introduced to our new supervisor, TSgt Frankie Howell. A Mississippian who grew up dirt-poor on a cotton farm, he made it crystal clear he didn't like white people and definitely did not like me. I didn't know what to say, so I just shrugged and replied, "OK." That was the wrong response, apparently. He said he had heard all about me and that I had to earn his trust. Ron tried to speak, and he snapped, "Shut the hell up. I know you two are friends, and I don't like your black ass either." So off we went to our respective gates.

I was posted at the North Gate by the base hospital. Shortly after, TSgt White showed up to check on me. We talked for a while, and he asked if I had any questions. I asked if he knew why TSgt Howell seemed to have it in for me. He replied, "It's not personal, and don't take it that way. He grew up poor and has had to work hard for everything he's got. He'll warm up to you with time. Just be yourself, do your job well, and things will take care of themselves."

He started to walk away to his patrol car. I asked if he'd heard what was said about me from the Philippines. He smiled and said, "Look, you came here with a lot of experience and know a lot more than your records reflect. As far as I'm concerned, anything not in your records is hearsay. A lot of shit happens over there and you made some enemies. You have a clean slate with me." He winked and got into his patrol car, adding, "Don't worry about Frankie. Just do your job well."

Depending on personnel turnovers, leaves, and school absences, we averaged 15–16 LEs for duty in any given 9-day cycle. In addition to TSgts White and Howell, we had two K9 guys — Sgt Joe Lyons and SrA Bobby Goggin. A1Cs Larry Splawn (Waxahachie, Texas) and Sidney Wright (Detroit) were our desk sergeants. Everyone else (Rucker, SrA Bill Upchurch, SrA Gill, A1C Bailey, Airmen Greg Sly, Gerald "Bogie" Boggess, and a few others) split their time between gates and patrols. Splawn soon transferred to another LE shift, so Ron Rucker replaced him on the desk.

My former 3rd SPG colleagues Don Adams and Mark Waters were also on our flight. Don was from Irving and his parents' house was near Cowboys Stadium. Mark was from nearby Hurst and his dad was a retired E9. However, I was greatly surprised to discover Mark was still an E3 and not an E4. I soon learned he had been promoted and then demoted due to a feud with SSgt Tony Schneider, his supervisor on day shift LE. To separate them, Mark was reassigned to D Flight LE pending an early discharge scheduled for a few months later. Those with knowledge of the situation felt Mark was treated badly by Schneider and reacted as one might expect.

I soon had a contentious encounter of my own with SSgt Schneider. He was a bitter and pessimistic little man with an oversized ego. I checked two antique firearms (a 12-gauge shotgun and a .22 rifle left to me by my late grandfather) into the base armory until I moved off base to an apartment. It was routine for day shift desk sergeants to check serial numbers of personal firearms through the National Crime Information Center (NCIC); however, SSgt Schneider took things too far. When the NCIC check on my firearms came back "negative," he assumed they were illegally owned since the serial numbers were not registered. In reality, the NCIC database didn't track legally owned firearms or antique arms; it only tracked those reported stolen or used in criminal activities.

So Schneider called me on my gate one afternoon to inform me that the armory was turning the weapons over to civilian authorities; he said I was guilty of violating military law and would face questioning over possessing them. Beyond the assumptive stupidity of his accusations, I didn't appreciate his condescending tone of voice. I went from "0 to 60" in seconds and we got into a nasty argument over the phone. He threatened to have me charged with insubordination and other charges. I told him to feel free, that this would put me in front of the SJA where we could sort this crap out. He started yelling again, and I replied, "You either send someone to read me my rights or shut the hell up." I hung up on him, and 30 minutes later Captain Neal appeared at the gate to ask for my side of the story. He promised to run this down and get back to me. Neal seemed pretty confident Schneider was trying to pole-vault over a mouse turd.

The next day was our 24-hour break and we were at the 7th SPS building that morning for training. I got called in to see the executive officer (a major whose name escapes me) to report where I obtained my firearms. I told him my grandfather bought them new in the 1930s and that serial numbers for older firearms were not in the NCIC database. I said that SSgt Schneider had no business threatening to seize legally owned property without first coordinating with SP Investigations and the Staff Judge Advocate. I asked why Schneider felt I did not warrant legal due process as required by military law and the U.S. Constitution.

The XO jotted down a couple notes and said he would make a couple calls. I went back to my training class and he caught me as we broke for lunch. He said, "Airman Clark, you're clear, and this was a big misunderstanding. Don't worry about it. I'll deal with SSgt Schneider and his threatening to seize your firearms without authorization. That falls within our area, not his. Also, you need to avoid any future confrontations with him." Captain Neal was standing next to me as the XO spoke, so I had a witness. Neal advised me to check my firearms out of the armory and keep them at someone else's place until I moved off base. So I checked them out, and Ron Rucker let me keep them at his apartment.

The morning after our first midnight shift, SSgt Schneider drove the patrol car that brought my day shift relief out to my gate. I didn't know why he wasn't working the LE desk and didn't ask. We rode in silence until we reached the armory parking lot. As I opened the door to get out, he said, "You need to learn some respect for your superiors, Clark."

I snarled back, "And you need to learn common sense restraint and how to follow proper legal procedures, Schneider." I slammed the door before he could respond and walked away to turn in my sidearm. Captain Neal and TSgt White were at the clearing barrel, and White asked me what Schneider said to me. I replied, "Oh, he was telling me all about how he enjoyed older boys sexually molesting him as a child." Neither laughed, but the captain's faint grin spoke volumes. SSgt Schneider never bothered me again, and before long he transferred overseas. I discovered what many in the Philippines warned me about: those who had only served in stateside assignments were typically overbearing and hostile to everyone.

Tammy Scheff was supposed to report in October 1982 from the Philippines but was in a motorcycle accident the week before rotating stateside. She finally arrived after Christmas following home leave to Long Island, New York. She was still on crutches when I picked her up at DFW Airport. We always got along well because she was a solid patrolman and good at her job. Although some asked if we had dated,

Tammy and I were never more than work friends. I never dated work colleagues due to job risks and conflicts of interest, plus I had never seen anything good come from co-workers dating. Those relationships almost always ended badly and eroded team morale.

Several other NCOs and airmen joined Delta LE over time, including TSgt Tom Child, SSgt Roger Gillinger, SrA Bruce "Max" Crosby, plus Airmen Kenneth Price, Ken "Guru" Ruh, Ross "Roscoe" Wood, Marsha Gillihan, Vernoy, and several whose names I can't recall now. Two of the latter were new female airmen who seemed to have nonstop "guy problems" but never to the extent that it affected their duties or relations with peers. SSgt Mark Chaires later replaced Joe Lyons as our K9 handler. Mark, Ron, and I often went fishing or shooting on our days off. Airman Todd Swerske joined us after K9 school. SGT Jerry Haller joined us later on; he was a big pro baseball fan who was happy to be back stateside and near a pro team (Texas Rangers). A1C Gerald "Ghandi" Smith then replaced Chaires, who transferred to the Philippines. When Swerske transferred to Korea, we picked up SrA "Super Dave" Lynch as his K9 handler replacement.

198104: My friend Airman Bruce Bielby standing in the engine cowling of a C-5 Galaxy in Apr 1981. If caught, he would have lost a stripe for this stunt.

Several of the new airmen were a challenge and got in trouble both on and off duty. One whom I supervised was a heavy smoker and heavyset; he looked sloppy in uniform and whined about everything. He liked to sit in the gate shacks instead of standing the post as directed. It was both dangerous and rude to have a cigarette in hand while checking IDs, so that created more problems for the lad. Exacerbating his penchant for underage drinking, he liked to run his mouth and pick fights at the barracks. He then got in bigger trouble for setting his bed on fire while smoking in bed. Our captain and shift chief told me to "handle it," but the kid wouldn't listen to reason. I finally threatened to get several NCOs together and give him a serious "wall-to-wall counseling" (beating) if he kept doing stupid stuff. He dodged this by transferring to another shift and then requested a remote overseas assignment. We guessed he matured and did OK overseas since we never heard from him again.

Our flight was ethnically diverse (white, black, Latino) and worked well together, sort of our own little tribe. The only one most never got along with was Goggin, a Yankee with a personality so abrasive that he was alternately called "Baby" and "Masshole." This lad was presumptuous, eternally irritable, and nasty toward those of greater rank or higher intellect. He also talked trash about Southerners — ill-advised considering his peer set. He left active duty around my one-year mark on Delta flight and was missed by no one. The guy was a legend in his own mind and had no real friends.

Other than the SPs, I knew a few others from the Philippines — both assigned to the base hospital. One was Mike Bosarge, a Cajun X-ray tech who was friendly but drove home to Louisiana a lot. The other was Lisa "Rikki" Richards, an OB/GYN assistant from Roanoke, Virginia. Lisa and I became close friends (purely platonic) and hung out a great deal over the next two years. She was like a big sister to me and we went places a lot with an extended group of friends. She married a younger airman shortly before I left the Air Force and we lost contact for years until reconnecting after we both retired. One Security guy who arrived from the Philippines was named Youngblood; I seem to recall he extended a year there, so was only with the 7th SPS for a year before getting out to attend college.

As soon as I got on D Flight, before Don Adams even said hello, he told me, "You're going to hate it here. It's worse than the P.I." He explained his reasoning, and I told him we'd work together to change it for the better. Don was working as our LE desk assistant and became good buddies with Ron Rucker, so they were soon our desk sergeant team. They were good at running the LE desk and dispatching, yet they cut up a little too much at times. They persisted in telling ethnic and dick jokes, making them equal-opportunity offenders to every race, religion, sexual orientation, and nationality. While most either ignored them or didn't care, TSgt Howell stayed hopping mad at them (especially Don) over their jokes. When Ron protested that he could tell black jokes because he was black, Howell corrected him by saying, "That doesn't give you a bye!" During that exchange, Don Adams nearly caused Howell to have an aneurysm by asking, "So does that mean we can still tell redneck, Mexican, and blonde jokes?"

TSgt Howell clearly had a "thing" for Bailey, and we all suspected they had a secret relationship. While I'm fairly sure they did not, multiple indicators suggested it. Far junior in rank and much less experienced than me, she worked triple the patrols I did. She frequently rode with Frankie, and she often wiped sleep from her eyes as they pulled up to a gate whenever I was relieved for midnight chow. I never mentioned it to TSgt Howell and honestly sensed he dared me to say something. While I did mention it to TSgt White, he was retiring soon and didn't want to get involved. He just brushed it off and reiterated that Howell was in charge of the duty roster and schedules. Evidently, he didn't want to challenge his flight chief heir. Everyone else noticed Bailey constantly riding with Howell, so her unofficial nickname was "Sleeping Booty." She detested the name and we didn't give a damn. As Don Adams told her, "The glass slipper fits, Cinderella, so you get to wear it." We also called her "The Princess."

One night while assigned with me at the NMSA, Sleeping Booty ignored my order to stay awake during alarm panel duty and took a deep nap, so I taught her a lesson. Don was our third man that shift, and while we were out on patrol, I had him kick an igloo door to trip the alarm. She couldn't make radio contact with us, so she assumed we were under attack and called the base for a QRF. Captain Neal called me on the radio and I reported all was well. He drove out there anyway and asked me what was going on, so I told him. I added that we needed to teach this airman a desperately needed lesson in following orders and staying alert on duty before she got someone killed in a real incident. The captain concurred with my assessment and drove back to base without confronting her over it.

I worked primarily gates or the NMSA for most of the next six months. It was hot and humid until late October. The swings were busy until around 1800; by then, everything on base was closed except the chow hall and the Officers' and NCO Clubs. The Marina desk tended to stay open until midnight or later due to boaters out fishing late. Unlike in the Philippines, the mids were quiet and boring. On rare occasions, incidents erupted at the clubs or at a barracks, but usually things were very tame. We got maybe one actual criminal incident per cycle—usually minor stuff—and I prayed to God it happened on my shift. With greater patrol experience than my peers, I preferred to work patrols but was OK with working the gates, especially on mids.

TSgt Howell always grew hostile when I mentioned posting assignments, so I left it alone.

The short time I lived in the barracks was uneventful. It was noisy at times from bored off-duty guys getting bombed. Most were good about keeping stereos low, but a few weren't. The dayroom had a TV and it seemed to stay on MTV all the time. In those days half the military smoked and there was no prohibition against smoking indoors. The dayroom was a dragon's den of cigarette smoke and stayed trashed as a result. One night I walked in after a swing to find a drunk airman passed out on top of the two side-by-side soft drink machines. I punched him in the nuts for being stupid and he groaned but never woke up.

During late afternoons, when the Delta Security guys in the barracks woke up after mid-shifts, they gathered in the shade on the east stair landing and hung out. Mark Spinney, Cave, Diamond, Grant, Streeter, and a big kid nicknamed "King George" (so named for a freakishly large anatomical feature he enjoyed showing off) were the ones I remember most, mainly because they were hilarious. Grant's roommate, Streeter, neither brushed his teeth nor changed his sheets on Fridays — good guy, just poor personal hygiene. Anyway, the Deltas would gather on the metal stairs to tell jokes, sing nasty songs, crank up heavy metal on their ghetto blasters, and yell at people driving down the barracks row road. During off-duty evenings, they headed into town and got really rowdy. These guys stuck to each other like wet summer dog turds. Several dozen liked to descend on small honky-tonks during "two-for-one" nights and try to drink the bars dry. While I didn't run around with them, we always got along well. Six months after I arrived, a half-dozen of the Delta Security guys got nailed by a urinalysis for smoking dope and were kicked out of the Air Force.

One of our married guys caught a tarantula at his house and gave it to me as a pet. I kept it in a big jar in my barracks room and fed it insects. Bugs typically saw the tarantula and went wild running around the jar before the predator sprang on its prey. An unplanned benefit was that an airman across the hall was frightened by spiders. One time, the dickhead walked into my room and grabbed a beer from my fridge without asking. We got him down and let the tarantula crawl all over his belly while he screamed his apologies. He never set foot in my room again. I kept the tarantula until the next spring and then set it free.

I pinned on E4, Senior Airman (SrA), on 1 October and moved out of the barracks to a furnished efficiency apartment. It was in a new complex about six miles southwest of the base in an area recently developed from ranch land. I paid $245 per month, utilities included. The complex owner was a WWII and Korean War veteran (U.S. Army infantryman), so he liked having active-duty people living in the complex. Working shifts, I never got to know any of my neighbors. With the place under the base's flight pattern, it was difficult to sleep at times during the day due to B-52s flying overhead.

In early November, I failed a no-notice standboard evaluation. The Training Section managed and ran the program independent of the SP flights. I passed the written test easily but missed a key procedure on the practical exam; I simply omitted one step, which caused an automatic failure. TSgt White didn't say much, but Captain Neal chewed my tail royally. It stung to have let them down. TSgt Howell just glared at me, but I really didn't care what he thought. A month later, I received a no-notice makeup evaluation and scored a perfect 100%. Thereafter, I had similar standboard exams regularly and scored 100% again the next year too.

I flew home on leave during Thanksgiving week. Early in the week, I drove up to Bowling Green to see Hannah, now a sophomore at Western Kentucky University. She mentioned I had some "competition," but insisted I was her only real interest. While I had met and casually dated a beautiful Tejana girl in Fort Worth, I refused to let it grow beyond friendship because of Hannah. Her revelation of my "competing" for her angered me, so I made it clear I wouldn't wait for her to figure it out. Hannah asked me to give her time to work through it and I naively acquiesced.

Uncle Zollie Webb always had a big family reunion at his farm down Fall River Road, so I accompanied my parents to Thanksgiving there. We also drove down to New Hope, Alabama, to visit my dad's family. I was surprised to see a photo of me in uniform on my 95-year-old great-grandmother's nightstand. After a couple more days of hunting and doing things around the farm, I flew back to Texas.

When taking leave, the 7th SPS did not require us to sign out on the first day and sign in on the final day. Instead, we were expected to show up for work on time after our leaves ended. A shift worker could sign up for six days of leave to cover three swings and three mids, yet still take advantage of the three days off prior to the leave and the three days off after it. Whereas guys were "officially" approved only for six-day absences, in reality, most were gone for 10–12 days. You took a chance when traveling out of state, so it was wise to schedule return flights two days early to accommodate potential travel delays. Since we accrued 2.5 days of leave per month, I normally took a six-day leave every four months and flew home to Tennessee. By taking only six days at a time, I "banked" an extra 12 days of leave per year that I could use for terminal leave later on.

MSgt Reggie Phillips asked me to try out for the EST in December. He knew I had been an EST member in the Philippines and was rated expert on every weapon in our armory. During the tryout, I did 65 push-ups and 66 sit-ups in the 1-minute drills, then ran a 5:24 mile. The 50-foot rope climb in full gear followed, and I climbed it without a problem. Phillips assigned me to the squadron team without caveat, and I was glad to join.

As part of the combined squadron team, each flight had 3–6 EST members, and I started working out with ours. We trained for two hours after the first two swings and sometimes during the 24-hour breaks in each 9-day cycle. Because the time commitments didn't sit well with wives, most of our members were single or divorced. MSgt Phillips often trained with us and rotated me through different positions. After several months, he announced that he was nominating me and Sgt Rick Geiger to attend the 3-week EST/SRT Team Leader's Course at Lackland AFB.

198009: Basic training photo, Lackland AFB, Sep 1980

Most EST training scenarios were designed to make us "lose," because we learned more from losses and draws than we did from wins. We always conducted detailed AARs to determine what went right, what went wrong, and what lessons we could take forward and practice again. Some nights, we might train multiple times for two straight hours on a single scenario just to refine our response, maneuver, and engagement drills.

I had my first NCO professional development course in December 1982 — very easy stuff. A basic training buddy from Guam, SrA Velez, was also in my class. While we had a surprise four-inch snow the morning of graduation, the ground was warm, so the streets were never slick.

I didn't go home for Christmas and worked both Christmas and New Year's Eve. I wasn't dating anyone and had nowhere to go, so I spent my off weekend with Ron Rucker's family. I liked his wife, who somewhat resembled Princess Diana, and she was always very nice to me. I enjoyed visiting with them but got annoyed with Ron Jr. one time for jamming several oatmeal cookies into my Betamax that I had loaned them; fortunately, we were able to shake and vacuum all the pieces out. While I sensed something wasn't right between Ron and his wife when she first arrived, their problems soon surfaced, and a year or so later they divorced.

198201: On New Year's Eve 1981, SSgt Quesnell and I were tasked to serve as body guards for Philippine pop singer Imelda Papin at the NCO Club. Because she was extremely popular and liked to mix with the crowds, our mission was to ensure no one grabbed her or snatched jewelry from her. It was a great evening and the singer returned to Manila safe and sound.

Late one evening when we were on our 3-day break, I received a call from Don Adams requesting that I come get him at Diamond Jim's, a honky-tonk near my apartment in Western Hills. He sounded trashed and soon a policeman took the phone. He said Don was drunk and got into it with some other patrons and that the only reason he wasn't already under arrest was because he was an SP. I thanked the officer and promised to immediately come retrieve him. I called Ron Rucker to assist and picked him up on the way.

We arrived at the honky-tonk and were not surprised to discover Don handcuffed to the porch railing outside. I located the cop so he could remove his handcuffs after I put mine on Don. While Don was pretty upset that I handcuffed him, there was no way I could chance this big fellow getting crazy on me. We stuffed and buckled him into my front passenger seat, and Ron drove my car while I drove Don's truck. I asked him, "Are we taking you to your apartment or to your parents' place?" He chose his apartment. We got him home and tucked into bed, then hid his truck keys and left. As I dropped off Ron, he remarked that he was glad Don didn't turn on us. I agreed, because a drunk, angry man Don's size would have taken both of us to subdue.

The next day, Don called me wanting to know what happened, how he got home, and where his truck keys were. He didn't remember a thing about the night before. He apologized profusely, thanked me for taking care of him, and promised never to go out drinking alone again.

From then on, a bunch of us went out together like a posse and always had a designated driver. Typically, I got tagged as the DD. I didn't mind at all since most places gave DDs free food and non-alcoholic drinks. I also charged my riders a buck or two each, depending on driving distances. Gas was relatively cheap ($1.25 to $1.50 or so per gallon), so I often filled my tank on my DD fees alone.

Coach Paul "Bear" Bryant died in late January, and it made

national news. His passing was very big news in Texas since he had coached at Texas A&M for a few years in the 1950s. I was on a patrol and tasked with raising the Colors at Wing HQ the next morning. Rucker told me during the wee hours to put the Flag at half-mast at 0600. When I asked who died, he said he didn't know and then added, "Maybe it's for that Alabama coach, Bear Bryant." So, hearing nothing else, I figured that was the reason.

When I radioed in at 0605 that the Colors were posted, I said, "Roll Tide!" Someone radioed back, "Long live the Bear." Others radioed, "Roll Tide!" or "Go Bama!" and similar remarks. This went on for a while from both our guys and the Security grunts out on the flightline. We didn't realize the Wing Commander, Colonel O.K. Lewis, was monitoring our frequency. He suddenly came on and told us to be quiet, informing us that a retired general officer had passed away. However, in our minds, the Flag at half-mast was still for Bear Bryant.

While it snowed lightly only a couple of times that winter, it still seemed long and cold. Naturally, on the coldest nights, I was assigned to gates while those junior to me— "Sleeping Booty" and the other females—rode around in warm patrol cars. Frankie Howell saw to that after taking over from TSgt White, who retired before Christmas. Ron Rucker and Don Adams were our LE desk sergeants. Sometime that winter or spring, SSgt Roger Gillinger reported in as the deputy flight chief and patrol supervisor; he had served as a desk sergeant before and understood every aspect. I could tell right away that Gillinger wasn't fond of Howell, yet he wisely steered clear of the ongoing drama.

Roger was by the book and no-nonsense, plus he was our flight training NCO. While I greatly respected him, it took me a couple of months to earn his trust. What gained it was my diligence with building checks; both he and I spot-checked behind other patrolmen to ensure they physically checked every door. Bravo Flight had mids behind our swing shifts and rarely ever found an unsecured building. However, Charlie Flight had swings ahead of our mids, and for months we found a disproportionate number of unsecured buildings. We didn't know Charlie Flight was under investigation for stealing government property, and OSI was building a case that would put a lot of SPs behind bars.

For some reason, most peers hated pulling duty at the off-base NMSA. I asked to have it twice per 9-day cycle and usually did. I loved NMSA duty. It was the original base nuclear dump in the early Cold War era, located seven miles west of our base (10 miles by road) out on White Settlement Road. It now held only "dumb" bombs, sea mines, 20MM rounds, .50 caliber ammo, and tons of various small-arms ammo. It had a guardhouse with alarm terminals next to the entrance sally port, a dozen bomb igloos, a water tower, two main munitions buildings, and three smaller buildings. The NMSA was roughly 600 meters long by 400 meters wide.

198308: Senior Airman Dan Clark at the off-base bomb dump or NMSA in August 1983

We posted a three-man team at the NMSA during swings and mids. One stayed with the ancient drop-flag alarm panels while the other two patrolled or hung out in the guardhouse. The NMSA had an old beater pickup — a rolling heap of rattling, squeaking parts — that served as our interior patrol vehicle. Other than driving around to manually turn on the big area lights at dusk and then off at dawn, there was little to do other than PT, study professional materials, read, or just relax in the guardhouse. Surrounded by ranch land, it was quiet out there under the big starry sky, and no one bothered us for the most part. Occasionally, fraternities from Texas Christian University sent pledges to attempt setting off the alarms — either by scaling over the back fence or shooting at bunker doors from afar. That only happened once on my watch, and we expected it after several trucks stopped on the access road. Alarms went off two other times, but they were only false alarms.

The NMSA lights drew insects (especially crickets and grasshoppers) by the zillions, and in turn, this lured in many four-legged varmints to feed on them. The latter were mostly armadillos, skunks, raccoons, and possums. I often sneaked my personal pistols out there to hunt after the munitions guys went home. Usually, I just went "Dukes of Hazzard" and ran the buggers down with the old truck. One mid-shift I ran over 8 skunks, 2 possums, 3 coons, and 2 armadillos; if I recall correctly, that was the record for a long time. I reasoned that buzzards and ants needed to eat too. While most peers were cool with it and joined in, we just omitted those who weren't from our activities. Anyone I viewed as untrustworthy or too sensitive stayed in the guardhouse with the alarms.

The views from the water tower were great. Whenever I was posted with two LEs who wanted to lounge in the guardhouse all night, I could climb up the tower and enjoy the scenery. Seriously into photography at the time, I took some decent photos of Fort Worth's night skyline off in the distance. I could also shoot at varmints from the tower and ground with my .22 Ruger pistol and bagged quite a few. One night I hauled my M16 up the tower with me and picked off a coyote beyond the perimeter fence. I always brought my own 5.56MM rounds and a can of carburetor cleaner, then cleaned and oiled my rifle thoroughly before heading back to the base. The armorers noticed that I regularly cleaned my rifle while posted at the NMSA, thus my handing them a freshly oiled rifle was nothing out of the ordinary.

At one point, Captain Neal was TDY for a couple of months to Squadron Officer School at Maxwell AFB, Alabama. A first lieutenant temporarily replaced him, and he was a fairly laid-back fellow, a prior service NCO who returned to active duty after college. One night I was the senior LE out at the NMSA and had already run down several skunks and armadillos when the lieutenant showed up unannounced for a post check. As we drove around, I explained our operating procedures; I had these thoroughly memorized, so I didn't require notes. He started chuckling, and I asked him what was so funny. He said, "Now tell me about all your extracurricular activities."

I asked, "Excuse me, sir?"

He replied, "This place smells like skunk hell, Clark. I know what y'all do out here." I didn't know where to start, so he offered, "Show me your victims. I've heard all the rumors, so I need to see this for myself. Take me to 'em." So I drove past each one, and he nodded out the window at them without saying anything.

The southeastern munitions maintenance building had a large concrete parking area out front. I had run over a large skunk on the concrete, and it sprayed a big puddle during its death throes. The lieutenant grinned and observed, "No wonder the maintenance guys bitch and whine so much about y'all. Do you do this every night?"

I answered, "Yes, sir. It helps us stay awake and alert." By then he had seen enough and told me to take him back to the guardhouse.

As we left the concrete pad, the headlights illuminated an armadillo waddling across the paved road. The lieutenant said, "You might as well get that one too." So I floored the gas, swerved left into the grass, and crunched it. He was still laughing as he got into his patrol car to return to base. If he told anyone about our little adventure, they never mentioned it. This was just one of those loony aspects of military life we didn't talk about.

The next morning, as I was turning off the big area lights at each pole, I passed by the munitions maintenance building where I had run over the large skunk. An airman was out front, pushing the skunk across the concrete pad toward the grass with a wooden 2x4. As I passed, he yelled, "Hey asshole! Thanks for running over this f—king skunk last night and leaving it for me to clean up!"

I asked, "What makes you think I ran over it?"

He yelled back, "What the hell else could have killed it?"

I replied, "I don't know. Maybe natural selection?"

He bellowed, "F—k you, asshole!"

I laughed and waved, "Not a chance! Have a nice time with your new pet!"

When I got back to the guardhouse, our day shift relief was already there. The patrol leader was a new sergeant who had never before worked at the NMSA. He frowned and observed, "Man, this place smells like skunks, cut grass, and cow shit."

I said, "Yeah, it sure does. Welcome to Texas."

I flew home on leave again in February 1983. I drove up to see Hannah at WKU but did not enjoy my visit. She admitted to still dating the old family friend but said it wasn't serious; she kept insisting I was "the one" and that she was simply lonely with me so far away. Despite being very hurt and upset over her revelation, I stupidly believed her and elected not to break it off — at least not with 17 months to go on my enlistment. I spent the rest of my week home visiting family and friends, fishing, and going to my brother's basketball games. I wouldn't get home again until the following August.

Late in the winter during a cold snap, TSgt Howell slipped on some ice and fell down his apartment's concrete stairs, suffering a closed-head injury. He almost died and required brain surgery. While I hated to see it happen, I was very happy to see him leave Delta. Disqualified from carrying a weapon again, after recovering he was reclassified out of LE and assigned to administrative duties. I very rarely saw him again and was nice to him when I did. I felt great pity toward him because, after the injury, he seemed like a shadow of his former self.

MSgt Noble Smith was reassigned as our flight chief and brought TSgt Tom Childs with him as his deputy. With Ron Rucker transitioning over to SP Investigations, SSgt Gillinger took over as desk sergeant along with a new guy, SrA Bruce "Max" Crosby. Bruce had served as a maximum-security guard at a corrections facility in Germany. The unexpected aspect of our personnel reassignments was Gillinger's recommendation that I move up to patrol supervisor. This surprised me since several others were ahead of me by rank or time in grade; however, they were all either getting out soon, had transfer orders, or didn't want the responsibility. I accepted the duty and was determined to excel.

The one person who gave me crap was SrA Goggin. Although the same rank, he was extremely jealous of me and everyone else. I was on leave when Delta Flight conducted the annual physical training test, and MSgt Smith promised a 3-day pass to the fastest runner on the 1.5-mile run. Goggin ran it in 10:37 and pestered Smith to declare him the winner. Smith refused since I had not yet taken it, but Goggin kept bugging him. Smith finally said he could take the pass if I ran a slower time when I returned. Most Delta LE guys knew I was working out like a dog with EST and running the mile in the 5:30 range, so they laughed at Goggin for his naïveté. When I returned from leave the next week, I took the PT test and ran the 1.5-mile in 8:42. Goggin was furious since he was ineligible for the pass he so desperately wanted.

During guardmount the next evening, Goggin badmouthed me before formation. Then, during it, he openly demanded MSgt Smith not give me the pass. I remained silent until I finally had enough. I turned around and asked, "Airman Goggin, do you have a personality disorder? Enough already!" As I turned back around, Goggin flew into a cursing tirade that drew Captain Neal's attention. MSgt Smith ended it by telling Goggin to be quiet, then said he would award us both a pass. When I volunteered to let Goggin have my pass to keep the peace, MSgt Smith shook his head no, adding, "You actually earned yours." This made Goggin even madder. I was heading to the NMSA for the mid-shift, so I didn't see him again that night.

The next morning, Goggin tried his best to pick a fight as we turned in weapons. I ignored him and went about my business, which goaded him so much that he followed me to the clearing barrel, still yelling at me. Spit flew from his mouth as he raged like he had rabies or worse. Captain Neal took notice and walked over; he ordered Goggin to be quiet and disappear before he charged him with disorderly conduct. MSgt Smith also revoked his 3-day pass for this outburst. I declined to take my own pass as penance for my role in this saga.

Losing his cool in front of so many people destroyed what little respect anyone had for Goggin. From then on, until he left active duty a month or so later, he kept to himself and steered clear of me. Several guys threw him a farewell party that was reportedly sparsely attended by anyone from our shift. Goggin stayed in the local area for a short while before he reportedly migrated back to Massachusetts. Reflecting on it later, Mark Spinney (a hilarious and popular Delta Security guy from Boston) said he never understood the man, that he certainly didn't represent his home state in a positive light.

A great guy named Gerald Smith replaced Goggin. An older airman, he was married, a Mormon, and a calm professional with no vices. We nicknamed him Gandhi because he looked a little like Ben Kingsley, who had played Gandhi in the movie of the same name. Everyone loved Gerald and he brought balance to our LE team. We stayed in touch for years and he made the Air Force a career, eventually cross-training into information systems management.

I moved in with Don Adams in May when his previous two roommates moved out. His apartment was in a sprawling, older brownstone complex behind the Ridglea Theater south of Camp Bowie Boulevard. The residents were mostly elderly people and Tejanos. We had a third bedroom and lifted weights in it twice a day. Don had a great gig moonlighting as a security guard at the apartments, so he paid only $100 per month rent. I was offered the same arrangement, so I was happy to help with security.

We never had many problems with the residents. Occasionally, someone called about a noise complaint or a wild party. We would go check it out and usually took care of the issue without having to call the Fort Worth Police for stronger intervention. With two swimming pools there, some calls were to run off kids from adjacent neighborhoods who sneaked in to swim. While some residents insisted we have them arrested, we refused to do that to young kids. Incidentally, we never had any problems from the illegal aliens living there; they tended to live quietly and under the radar to avoid deportation. Our biggest problems were first-floor parties that overflowed into the yards.

Our first apartment manager quit in August and moved to a home where she could spend more time with her kids. The new manager moved in right above our apartment and was a pain in the neck. She wanted one of us on the premises full-time and armed — not something we felt was practical or necessary. She also had a series of men visiting her and was noisy as hell when having sex. This went on at all hours, and finally we had enough and resigned our security duties. Don and I decided to take on another tenant for our third bedroom, which until that time had been our workout room.

Mark Waters moved in with us in September. He needed a place to stay for a few months until a vacancy opened at a condominium he wanted to lease. We had all served in the Philippines together and Mark had recently left active duty. He was from nearby Hurst and his father was a retired CMSgt. Mark had a job selling cars, so he always drove a new dealership loaner. He lived fast, partied hard, and frequently had women over. We went out shooting when we were all off, often at the ranch surrounding the NMSA or at the LBJ Grasslands up near Denton. Over time, North Texas experienced so much growth that Denton and the DFW Metroplex eventually grew together and merged.

Living with Mark was fine until he brought in Mickey — "The Kitten From Hell." It had a habit of pooping on the carpet and then attacked me one day as I slept after a mid-shift. It thought a tender part of my anatomy was a scratching post. Don already hated cats, so after the kitten attacked me, Mark took it to his dad's place. Don also complained about the women Mark brought home. Don had a girlfriend at Sheppard AFB in Wichita Falls, so he was possibly a little jealous of Mark having women over — especially if they helped themselves to his beer or food in the fridge. While Mark was a very clean and tidy roommate, he only lived with us for three months and moved out just before Christmas.

Don left active duty on terminal leave in December 1983 and moved back to his parents' place in January. Not long before he moved, I returned home from a swing shift to see him sitting on his waterbed fixing an appliance. I thought nothing of it and went to bed, only to be awakened later by a splashing sound. Near as I could figure, Don fell asleep and rolled over on a flathead screwdriver that punctured his waterbed. It drained rapidly with his big carcass lying on it, and eventually the water woke him up. He was floundering in the water, trying to extract himself without making a bigger mess or collapsing the water barrier sideboards holding the liner. I helped him out, and we spent hours bailing out water and pouring it into the bathtub. The next morning, he woke me up again running a wet vac cleaner he had rented to pull the spilled water from the carpet. I was too tired to get mad at him over it.

After Don moved out, he got a job as a trucker and stopped by a few times to visit. During one visit, he took my crock pot with him, promising to bring it back or pay me $20 for it. You know, that goon neither brought back my crock pot nor ever paid me for it. We've stayed in touch over the years, and the issue has become a running joke between us. Sometimes I message him and ask, "When am I going to get my $20 or the crock pot back?"

Due to the "6 on - 3 off" nine-day work cycle, I was off duty only one or two Fridays or Saturdays in any given month. During these times, I liked to go hit the dance clubs with my friends Lisa Richards and her roommate, Betsy Fowler. There were several big dance clubs in Fort Worth where the cover was low, the ventilation high, and the ambiance just right. I usually served as the designated driver and enjoyed the free non-alcoholic drinks (and sometimes food) that the clubs offered. Since I had a big Ford LTD with seatbelts for six people, we typically had four to six of us per outing.

One night, as I drove my friends (all of whom lived in the barracks) back to base, I saw a bright light ahead of me on I-820. I was in the left lane passing a slower car and had just returned to the center lane when a large, jacked-up pickup truck blew past me. The truck had obviously turned onto the wrong-way access ramp and was driving straight into oncoming traffic. I was going about 60 mph, and judging from the speed at which the truck passed us, it was going much faster. Betsy, who was sitting in the middle front seat next to me, said, "That truck would have killed us. I'm sure glad you moved over!" Everyone else was talking and didn't see the truck until it passed by. It was a scary moment for all of us.

One night Captain Neal told me to go pick up a deserter in Grand Prairie near Dallas. Our attached U.S. Army AWOL Apprehension Team normally retrieved deserters for the DFW area, but they were away training at Fort Hood. I took a big Security grunt named Stanley Caldwell with me as backup. Stan was on our EST, and no one in their right mind would want to tangle with him.

We picked the deserter up without incident and brought him back for in-processing. Bruce Crosby searched him and processed him in. Per SOP, he stayed in maximum security for 72 hours and afterward walked with a desk sergeant or rode with an LE patrolman to the chow hall. Visited several times by his parents and other family members, the deserter was cooperative and pled guilty to his desertion charge. The court was lenient, so it only imposed a six-month sentence with a general discharge, and he was allowed to serve his time at Carswell AFB before his release. He got along well with the desk sergeants and was allowed to perform various work details on minimum security around the base. Overall, he was a pleasant fellow and seemed like a decent guy who made an impulsive mistake.

Several months into his sentence, the deserter took off running while walking to the chow hall for lunch one weekend. He reached the eastern perimeter fence, climbed over, and was gone. It was a big embarrassment for Alpha Flight and the day-shift LE who lost him. Several weeks later, the Army AWOL team picked up the deserter at his parents' home near Dallas. Local police had put the home under surveillance, thinking he might show up or his parents might turn him in. When the AWOL team arrived, the guy was sitting in the living room watching TV, so he came along peacefully. This time the Air Force sent him to the military penitentiary at Fort Leavenworth, Kansas, for a much longer sentence. By the time he in-processed there, his original six-month sentence at Carswell would have almost ended.

One Saturday night I received a call to investigate a potential domestic disturbance at one of the barracks. I pulled in front of the building and it was strangely quiet outside until I heard intense yelling from an open third-floor window. I thought, Yeah, we have a problem. So I called for two backup units and within minutes they were on scene.

I sent one LE up the far-left outside stairwell and the other up the one on the far right, while I went inside and took the center stairs. We all arrived on the top floor almost simultaneously, but it was still quiet as I stood in the middle dayroom wondering where all the barracks occupants were. Then the yelling started again as a woman and a skinny guy emerged from one of the rooms to my right. At that moment, a big fellow emerged from a room on the opposite side and started in my direction, yelling at the woman. They continued to scream and curse at one another as they approached me in the dayroom.

As the big fellow entered the dayroom, I yelled for him to stop and move no closer to me. I turned and told the guy behind me to get the woman to shut up. About that time, the big guy grabbed the metal pay phone, ripped it out of the concrete block wall, and threw it sideways across the floor. Realizing I didn't want him grabbing me, I pulled my pistol and ordered, "On the floor — NOW — or I shoot!" He nodded and lay down. My other LE rushed up and handcuffed the guy, and seconds later the situation was calm again.

As expected, the big fellow and the woman were an estranged couple going through a divorce. He had moved to the barracks and, for whatever braindead reason, she showed up to talk to one of his co-workers. Of course, the husband blew a logic fuse, and it went south from there. I believe he received an Article 15 for his part in the disturbance and for destroying the pay phone. And I seem to recall the wife was barred from the installation. Both should have been charged with Grand Theft Oxygen.

In late spring 1983, MSgt Phillips informed me I was approved to attend the EST/SRT Team Leader Course at Lackland in July. A very stocky and muscled fellow, Rick Geiger, didn't make the cut due to weight, and we only had one slot. I trained hard on my own for the next two months and was given extra opportunities to fire before swing shifts at the base pistol range. I ran twice daily and got my mile time down into the 4:50–5:00 flat range. I was also lifting weights every day with Don and bulked up to nearly 200 pounds.

In early July, I packed my gear and drove to San Antonio for the first time since graduating from the SP Academy 31 months earlier. It hadn't changed much, and it felt strange to drive onto Lackland AFB. I stopped by my old basic training barracks and walked into the squadron area as if I owned the place. A drill instructor yelled at me before realizing I wasn't a trainee, then asked, "Oh, who are you looking for?" When I replied I wanted to see Sgt Brentham, he pointed toward the chow hall.

I walked in and saw now-SSgt Brentham sitting at a table sipping coffee with a couple of mess attendants. I stood over him, wondering what he might say. He looked up, recognized me, and asked, "Senior Airman now, huh? Are you here to kick my ass, Clark?" I started laughing, and he said, "Go get a cup of coffee and let's catch up." We sat and talked for a while before he had to go. As we parted, he said,

"I was harder on you than the rest because I saw talent. But you were a smart-ass and needed taken down a few notches. I knew you wouldn't take it personally and you didn't. So good luck in your course and I'll see you around the Air Force." He smiled and walked away with the familiar click-clack-click-clack of his drill taps echoing under the concrete and steel overhangs of the barracks.

I checked in with my training unit. While the three-week EST School was not a "gentleman's course," it was easier than I expected. I was well-prepared and ready for the physical challenges despite the oven-like July heat. There were 12 trainees in my class (3 officers and 9 enlisted), and since all other enlisted were NCOs, I was the most junior man. Two of the three lieutenants (2LT Police and 1LT Mosteller) were Army MPs, and there were a couple of MP NCOs. Our three instructors were seasoned SP NCOs. SSgt Lafferty was a 6'7" giant and didn't say much; he impressed me as easygoing despite his size. Since every class had a unique nickname, we chose Los Chicos Malos (The Bad Boys) at the suggestion of a Puerto Rican trainee named Sgt Solis.

Our training site was at the firing ranges near the perimeter with Kelly AFB. We arrived at the training site daily at first light, trained hard until lunch, took 2.5 hours off to eat and let our food settle, then trained again until evening chow. We ran everywhere we trained, all over the base. We lined up by threes on either side of our battering ram (a cross-tie with rope handholds) and carried it everywhere we went. Those not carrying the ram served as road guards to stop traffic at street intersections. Heat was a problem for everyone, including the instructors. I guzzled 2-3 gallons of water daily and still didn't pee much due to profuse sweating. We shifted to night training during our last week, and I was thankful for the cooler conditions. I was the point man on our team, responsible for recons (looking around corners and into windows) with my small mirror and reporting information back to our team leader. As point man, I was always the first through the door — "the fatal funnel" — during assaults.

I scored expert in all weapons events at the ranges, including the "stress firing" exercises, and also performed well in PT. During the final PT test, I did 67 pushups and 65 sit-ups in their respective one-minute drills and ran the mile in 4:44, the fastest run time in the class by almost two minutes. The instructors said my mile time set a new EST School record, so they were pretty pleased when the training squadron commander came to personally congratulate me.

Conversely, I struggled with some of the rope events. Hanging upside down on one tower exercise, I got dizzy in the heat and started to black out. I couldn't get turned around for the exfiltration climb — a major deficiency on my course evaluation — and my team had to physically pull me up. I also disliked a young lieutenant on my team, a newly commissioned Air Force Academy graduate. One night during an assault, he stepped on my right hand and I cursed out loud, not at him but in general; he got in my face and the instructors had to separate us. My temper was up, and I said things I shouldn't have until the senior instructor told me to shut up or he would send me home. I swallowed my pride and complied, yet still seethed over it. I was told later that the incident removed me from consideration for honor graduate.

All but one in our class graduated — an NCO who could not do pull-ups and was sent home the first week. He seemed lazy, so we didn't miss him. Our honor graduate was a TSgt who had the same overall score I did. He was a solid performer, and I liked him, so I was OK with it and glad for him. A photo of each EST class hangs in the SP Museum at Lackland AFB. While TDY to San Antonio in 1998, I stopped by the museum and saw it. Albeit strange, it was very cool to see my younger and skinnier self in the photo taken 15 years earlier.

While at EST School, I saw numerous friends from my SP Academy days and the Philippines. Among the latter group was Danny "Rabbit" Rogers (Jacksonville, Florida), with whom I hung out one weekend. He was a day-shift desk sergeant and off duty my first weekend. He remarked on how much bigger I had gotten since leaving the Philippines. We got together for evening chow several times too. During our second weekend off, I ran and worked out while the rest of the team hung out on base. I did go out with them clubbing that

Saturday night, and I found it a complete waste of time. Both Sundays I was there, I ignored the off-limits signs and ran the obstacle course twice through each morning. There was never anyone out there, and I enjoyed the quiet solitude in the mostly wooded area.

I lost over 20 pounds at EST School and got down to a wiry 175. When I got back to Fort Worth, my roommate Don was surprised at how "ripped" I was. I resumed weightlifting and gained back half the lost weight in a few weeks.

As soon as I returned to Carswell AFB, MSgt Phillips had a one-week local EST course scheduled. I helped him put together some of the classes. We did a lot of PT, and I discovered several guys were really out of shape. A few started referring to me as the "Marquis de EST" (sort of like the Marquis de Sade) for running the daylights out of them.

We did some of the training at Fort Wolters, a Texas Army National Guard post west of us near Mineral Wells. While firing one day, a helicopter circled and landed behind us in the range parking lot. A brigadier general got out with his pistol and asked, "You boys mind if I fire with y'all?" It was his range, so we were happy for him to join us. After a while, he thanked us and took off in a cloud of dust. Only in Texas!

Our training classes went well until the final day. During a climbing exercise at a fire training tower, one of our Delta Security guys slipped down a rope and broke his leg. I was the lead instructor, so I felt like crap. He wasn't upset with me, but I had to fill out a lot of paperwork over it.

A few days later, while I was sleeping after a mid-shift, my EST beeper went off around 1500. I grabbed my gear and took off for the base. Upon arriving at the armory, we were told it was an exercise; when our beepers sounded, we never knew if it was "live or Memorex."

The scenario: an unknown number of terrorists had taken over the B-52 Alert Crew Visitation Center next to the base marina. The terrorists had an alert crew and two families as hostages.

The center was a stand-alone building with doors and windows on all four sides. We drew up an entry plan and I posted the sniper teams. As we staged to approach the building, a large Black man walked out with his hands up, yelling, "I'm a hostage! I have to negotiate with your officer!" I and another EST guy moved up, grabbed him, and cuffed him. We ignored his pleas to take him to our officer in the staging area. Instead, we escorted him down by the lake for a body search. He had fake grenades in his pants pockets, so we handcuffed him to a tree and left him there. By now, he was really yelling and cursing at us for handcuffing him to a tree in the hot sun. We ignored him, thinking it was part of the exercise scenario. Either way, if he planned to set off "explosives," he would take no one else with him.

While I was away securing the "bad" hostage, the other of our two entry teams assaulted the center. We were called up and had just gotten inside when the exercise ended. We had a good AAR, and Captain Neal was pleased with our performance. As it turned out, the guy we cuffed to the tree was an SP major and one of the inspectors (oops). He had no ill feelings toward us and actually gave us high ratings for creative thinking.

I flew home for a week in August. Hannah was home from college for the summer, so I drove up to her hometown to see her. We had a good visit until "the other guy" showed up to say hello. I was angry beyond words and decided I had enough of this nonsense. I was sure Hannah sensed it, and her mother certainly did. As soon as I got back to the farm, I wrote her a farewell letter and mailed it before I flew back to Texas. Although I cared deeply for Hannah, it was liberating to end what any reasonable person would consider an unfair and very one-sided relationship.

The rest of my trip home was great, and I enjoyed visiting my family. When I got back to Texas, a letter from Hannah was waiting for me. I read it, then dropped it in the trash. I decided that once I was out in 11 months and enrolled at MTSU, I was amenable to it if she wanted to try again. For now, I was done and ready to date other women on a level playing field.

In early September, Captain Neal called me on the radio to meet in the BX parking lot. He got right to the point and asked me if I had ever stolen any U.S. Government property. I surprised him by stating I had indeed taken a few 20MM ammo cans from the NMSA to store EST gear. They were locked inside the storage trailer outside the Armory. I added that I had one 20MM can at my apartment containing more EST gear but that I could bring it back to base. Captain Neal asked if that was all I had ever taken, and I replied that it was. He got closer — maybe a foot away, staring me in the eyes — and asked again: "Are you absolutely certain that's all?" I repeated that it was and asked why. He seemed relieved and said, "Good. I already told OSI that I was certain you and our guys are all clean." He then explained, "Remember all those unsecured buildings you kept finding on mids? As we speak, OSI is arresting most of Charlie Flight LE and some of the Security guys for theft of U.S. Government property. A lot of people are going to jail for their involvement."

Indeed, I did recall all the unlocked doors and windows I found on midnight shifts. Neal explained that the Charlie guys found unsecured buildings and helped themselves to a few items at first, which turned into more items over time as greed and complacency set in. I also recalled Charlie guys throwing a lot of big parties with steaks and beer. According to the investigation results, they stole a lot of beer and liquor from the Class 6 store. They then figured out how to gain access to the Commissary and stole untold amounts of food. By running a thin rod or yardstick between the electronic doors and pushing downward on the pressure plates, the doors opened outwardly like magic. I had never noticed that these doors were not chained from the inside at night. They also stole electronic typewriters, telephones, office supplies, and God only knows how much else. That our own SP peers were stealing and pillaging the base of government property greatly angered me.

As it turned out, A1C Sandy Summers filed the initial complaint confirming the theft ring to OSI. The Charlie thieves tried to get her to go along, and she did—only to take detailed notes the entire time. OSI had already started an investigation, but Sandy's report filled in the missing pieces to the crime mosaic. Worse, guys who had gotten out a year before and lived locally were still attending "Charlie parties" and bragging about their exploits. Sandy turned over her notes to OSI, and after a months-long investigation the Charlies' house of cards came crashing down on their heads.

When the dust settled, Charlie's flight commander (a first lieutenant) was exonerated of involvement but discharged for the criminality committed by his people. Its flight chief, with 19 years of service, was sentenced to prison at Fort Leavenworth, and a bunch more were jailed or received dishonorable discharges. Since the seven-year statute of limitation for larceny had not expired, numerous former Charlie alumni were recalled to active duty and charged, so some of those dirtbags were jailed too. Those who were members of the USAF Reserve or Air National Guard were punished as well. Only three Charlie LE members survived this "Night of the Long Knives" ordeal: Summers and two new guys fresh from the SP Academy. Revisiting this more than four decades later, I may be off on some details, but this sad saga most definitely occurred and was a great embarrassment to our unit, our base, and the LE community as a whole.

Summers and I had served together in the Philippines and everything I knew about her was above board. However, she dressed and acted somewhat like a man, so most suspected she was a lesbian. While definitely suspected in the Philippines, sexual aberration was so rampant over there that no one cared about gay troops in the ranks provided they did their jobs well. Sandy was a good LE, and to her credit she never volunteered anything about her orientation. After the Charlie busts, they reassigned her to Delta Flight and put her under my supervision. I protected her to the extent possible, yet knew someone would eventually go after her for taking down so many LE peers. After I got out to attend college, Summers was gone within a few months. While persistent rumors suggested it was over her orientation, her sudden exit could have been a cover for her going to work for OSI or another three-letter Federal agency. I never found out and hope the latter was the case. I wish I knew where she is so I could tell her how much I appreciated her honesty and diligence.

A secondary effect of the Charlie busts was that the other shifts had to cough up LEs to rebuild Charlie Flight. We lost several guys and went to minimum posting until the base personnel office could transfer in replacements and new SP Academy graduates. A tertiary effect was that leave requests were sharply curtailed for the next six months. Having just split up with Hannah, I didn't plan to go home on leave again until spring anyway, so I was fine with staying put in Texas for a while.

In early September, I was summoned to see the base commander, Colonel Robertson. Figuring it was about one of the investigations, I was surprised to discover he wanted me to take over as the Carswell Honor Guard NCO in charge (NCOIC). In fact, he wanted me to start the next day since he had just relieved the current NCOIC for undisclosed reasons.

I went to LtCol Griffin's office and reported I was picked to head the base honor guard for the next 90–120 days. He already knew and told me to do a good job, that it was a lot of responsibility for someone so young. So that evening I worked my last swing with Delta Flight for a while. While Captain Neal was disappointed to lose me for at least three months, he said it was a great professional development opportunity. Everyone else gave me grief about having a sham job working day shifts.

The next day I reported to the 7th Services Squadron for my new duty. To my surprise, the outgoing NCOIC was still present and said he was training me for the rest of the week. The first thing we did was conduct a full equipment inventory. The honor guard had 24 old M1903 rifles for firing party duty; half were in bad shape and cannibalized for spare parts. It struck me as very odd that these fully operational .30-06 rifles were not stored properly in locked rifle racks—just shoved into a padlocked wall locker behind a single alarmed door. There were several full sets of color guard flags, plus star flags for each USAF general officer rank.

Although everything was accounted for, the weapons were very dirty and the supply room was in disarray as if a tornado had hit it. Consumable items (burial flags, blank ammo, gloves, and uniform items) were not properly ordered to predicted usage levels, so we were short of many items. Recordkeeping was a mess and nothing was filed. The outgoing NCOIC gave no reason other than he was too busy. He neither volunteered why he was relieved nor did I ask. I found out later the guy upset a bereaved family member on the phone over the scheduling of a funeral detail. Reportedly, other incidents had occurred and that one was the last straw.

That week I met with the Services Squadron leaders to receive an overview briefing of their operation. Colonel Robertson couldn't make the meeting and left word to come see him next week. We had three funerals that first week, so I went along as an observer and only took notes and a few photos. Procedurally everything was sound, and I was impressed with the team's performances. At week's end, the outgoing NCO signed over the keys and left, saying only, "Good luck, buddy. You'll need it."

The following Monday I called in the 45 honor guard members as a group to get to know them. Serving on the honor guard was an additional duty, so they only did this on call; however, this part-time duty looked great on evaluations. The 7th Transportation Squadron members doubled as our bus drivers, and we had three trumpet players for playing "Taps" at funerals. The 7th SPS and all the flying squadrons were exempted from providing honor guard members.

198309: Senior Airman Clark as Carswell Honor Guard NCOIC, September 1983

Although several team members were SSgts and Sgts who outranked me, I was the designated NCOIC and they were OK with it. When needed, I called selected team members, and they showed up ready to go. Each drew separate rations pay since honor guard duty frequently required missing meals at the chow hall. The Services Squadron paid to dry clean honor guard members' Class A/B uniforms—another very nice benefit.

The Services Squadron was commanded by a heavyset major on his way out. He was unfriendly and rarely had anything positive to say. Conversely, the squadron NCOIC, MSgt "Goody" Goodwin, was super nice and very professional. Among other things, the unit ran the Officers and NCO clubs and was responsible for the base janitorial and cleaning services, including linen exchange on barracks row. Goody shared horror stories about dirty linen they picked up from the barracks NCOs. He joked about cases where they preferred to handle the badly soiled sheets with pitchforks and take them straight to the dumpster instead of washing them. Even then, they still had trouble getting out blood and shit stains. I had already seen this from living in the barracks, so I knew he was right.

Realizing I needed help to straighten things out, I met with the major to propose bringing a couple of dependable airmen on board full time. He instantly non-concurred and told me there was "no way in hell" he would approve such a request. I asked if he minded me asking the base commander when I met with him. He laughed and said, "Go ahead. He won't approve it either."

I met with Colonel Robertson the next day. I laid out the same plan, insisting it was justifiable to reestablish procedures and create a first-rate honor guard team. He stared at me for a moment and said, "Pick any two people you want, and I'll approve it for 30 days. If I like what I see, I'll consider making one assistant permanent." Sharing that I had experienced resistance with my improvement plans, I asked if I could report directly to him for support requests, and he approved this too. Needless to say, while the major was shocked and very irritated with me, he could do nothing about it.

I asked the on-call SSgts which team members I should pick for the 30-day temporary duty. They recommended two really sharp airmen whose names I can't now recall. Their commanders were okay with loaning them to us for 30 days. They reported in the next day, and we immediately got to work. In between funerals and color guard functions, we got the supply room organized and the weapons cleaned by the end of my second week. I got Goody to order enough gloves, burial flags, and blank ammo to last at least six months based on average weekly usage. This ensured we had a basic load in case funding got tight, a supplier ran short, or consumables usage ran above projections.

I enjoyed cleaning and fixing up the M1903 rifles. These were classic, bolt-action .30-06 rifles our Soldiers used in WWI and early WWII. I pleasantly discovered that almost all the rifles tagged as "inoperable" only needed a complete disassembly and a thorough cleaning or a couple of replacement parts. I gave Goody a parts list, and he requisitioned the parts, plus let me take 10 rifles to a local gunsmith for refinishing and re-bluing. Within a few weeks, we had 22 of the 24 rifles in inspection shape and firing order. The remaining two rifles we set aside for spare parts until the new parts on back order arrived. By the time my temporary assignment ended, all 24 M1903 rifles were in firing order and looked good.

I pinned on Sergeant on 1 October 1983 — a very happy day. After a nice ceremony at the base theater late that afternoon, we went over to the NCO Club. Many senior NCOs (including some retirees) waited for us there and lined up in a gauntlet between the front door and the ballroom bar. Everyone wanted to "tack" the stripes onto the newest NCOs. A lot of eating, drinking, and singing followed, so the sober designated drivers got us home safely. I had a friend pick me up the next morning since my car spent the night at the NCO Club. After my welcome to the NCO corps, I completely understood ritual "beat-in" initiations by urban gangs.

During the three-plus months I ran the honor guard, we performed funerals almost daily. If they were in close proximity, we could pull up to three in a single day. During one 60-day period, we pulled 64 funerals and 20 color guard functions. Sometimes we had to split into two teams due to the distance between events. While funeral directors normally called to schedule ceremonies, occasionally families called us directly. I found that grieving families were harder to deal with than funeral homes. When booked solid, a funeral home could sometimes delay a ceremony by up to 24 hours to get us locked in for a military funeral. It was pretty common for us to coordinate with local police providing "blue light escorts" to get us through traffic or small towns so we could make it to cemeteries on time.

With 95,000 military retirees in the Dallas-Fort Worth "Metroplex" in the early 1980s, we stayed busy. Although Forts Sill and Hood were large posts with full-time honor guards, the Metroplex was out of their range to reasonably handle funerals. We arranged for Fort Sill to perform all military funeral details north of Denton, and Fort Hood agreed to cover all funerals and ceremonies in Waco and areas south. With so many retirees, Carswell had to do more funerals in exchange for much less geography. But the traffic was still a nightmare—ugh!

We did a military retiree burial way out on a ranch one afternoon. It was in the middle of nowhere, 100 miles away, including many miles down a dirt road. Our bus crossed more than a dozen cattle gaps getting there and drove by thousands of cows. Finally, we reached a gate to a family cemetery where a funeral home representative and some cowboy grave diggers met us. Although fall, it was still warm and we sweated in our uniforms. Accustomed to vehicles coming to feed and check on them, several hundred cows also joined us. They stood around mooing and staring, chewing their cuds, and swatting flies with their tails.

Before long a funeral procession approached and the bereaved took their places in the tent. Events went smoothly until the firing party fired the honorary volleys. The startled cows began running toward a low ridge and then circled back, bringing a cloud of dust with them that enveloped the entire cemetery and all the participants. As our bugler played taps, the cows joined in to create a deafening chorus. The grieving family members began laughing, including the widow. I had to yell the standard condolences to the widow as I presented the Flag to her. Through her tears, she smiled and said, "Sergeant, don't mind our cows. They're going to miss him too." It was good we didn't have another ceremony that afternoon, as we were covered with sweat and caked with cow-shit dust.

The Fort Worth Military Ball at the Ridglea Country Club off Camp Bowie Boulevard was a highlight of my service with the Carswell Honor Guard. Both U.S. Senators from Texas, numerous U.S. Congressmen, at least two dozen generals (including General Bennie Davis, the Strategic Air Command commander), and more officers than I could count attended. Many oilmen and wealthy businessmen were veterans and wore their service uniforms. The 8th Air Force Jazz Band from Barksdale AFB, Louisiana, was also there. Representative Jim Wright from Fort Worth was present before the event and told me that if we needed anything, we should see one of his aides. He thanked us and the band for coming to support this high-profile soiree.

198311: Sergeant Clark (right) during Carswell Honor Guard ceremony, November 1983

We were asked to present the Colors, stay and enjoy the event, then retire the Colors at the end. It was customary to get fed if we retired the Colors, and the food was great. Since the country club was only three blocks from my apartment, and knowing my guys might get loaded at the open bar, I parked my car there with one window rolled down an inch. That way my guys could slip in their dirty gloves and ascots for cleaning the next week.

We weren't allowed to sit with the crowd during dinner, so the club manager gave us a separate room where we could eat after posting the Colors. The 8th Air Force Band musicians shared our dining room, but only long enough to eat before returning to the stage for their performance. My guys were well-mannered until a retired three-star general walked in and handed us two bottles of whiskey. Three of my guys began sucking down the whiskey, and my team's behavior went straight to hell from there. A mock sword fight broke out, followed by one guy "walking the plank" off a table.

As the dancing started, Congressman Wilson walked in with two women on his arms and asked why we were sequestered in a side room. I told him we were waiting for the dance to end so we could retire the Colors. He said, "That's horseshit," and directed, "Y'all go dance with those ladies whose husbands either can't or won't oblige them." I had no idea who Charlie Wilson was until years later, when Hollywood made a movie about his assistance to the CIA's proxy war against the Soviets in Afghanistan. Tom Hanks played Charlie Wilson.

While I went to the latrine, my guys waded into the crowd asking women to dance. Returning to find them all gone, I spotted them dancing with a throng of elderly ladies who were cheering them on. I gave up and let them have their fun. Beyond that, I was too busy talking to the very cute first lieutenant leading the 8th Air Force Band. Although we danced a slow dance and I sensed a strong mutual attraction, she told me she wasn't allowed to date an enlisted man. She was honest, so I didn't take it as condescending. When I told her I was getting out for college in a few months, she laughed and said, "Well, come look me up when you do."

The time to retire the Colors finally came. Fortunately, most in the crowd were tanked up and half the attendees had already gone home, so no one noticed a couple of miscues from my color guard team. Two of our team members didn't drink, so they got the others home safely. I lived nearby and just walked home.

I soon got into it with the major about putting women on the color guard. He insisted that because women could not serve in direct combat roles, I could not put a woman on a color guard team. The sharpest female airman with whom I had ever served, Andrea Hillman, was on the honor guard. Tall and athletic, she looked as good as any man in a uniform. So I used her anyway and let the Equal Opportunity Officer at Wing HQ know for protective oversight. I also discussed it with CMSgt Chuck Jett, the 7th Bomb Wing Senior Enlisted Advisor. Chief Jett was an Alabama native whom I had gotten to know well. I used to stop at his house while on LE patrols to talk about Alabama football. He agreed with our case and briefed the wing and base commanders. Thereafter, the major never mentioned women on color guards and rarely spoke to me or anyone else. Barring the commission of a negligent or illegal act, he realized he had zero control over the honor guard anymore—plus he already had one foot out the door with his pending retirement. MSG Goodwin came in a couple of days later and told me Colonel Robertson approved using female airmen on our color guard teams.

One of our best team members, John Georgian, was a real wild man. He was the classic wise guy—smart, sharp in uniform, good at his job, razor-sharp wit—and we became friends. When he left the service to attend college, I put him in for an Air Force Commendation Medal that Colonel Robertson said he would approve if routed to him. John must have done something to anger his own unit commander because it was disapproved at that level. We lost touch for 25 years and finally reestablished contact in 2009.

We had one funeral incident that created quite a stir. After a funeral director scheduled a retiree's funeral, the late man's brother called me about the composition of the funeral detail. He beat around the bush for a while and then got to the point. He said, "I don't want any 'dark people' laying my dear brother to rest." He repeated it to ensure I understood and used the term "darkies" more than once, so I acknowledged his intent. His words really, really irritated me. I called every minority member on the team and explained why I wanted them on this funeral detail. As expected, all volunteered and were eager to go. The pallbearer and firing party teams were Black, Latino, and Asian, plus our bugler was Black. Georgian and I were the only two "non-darkies" on the team.

When we arrived at the cemetery, it didn't take long to figure out who had called me. One man glared at me the entire time and looked madder than a wet hen. The widow thanked me as I presented the U.S. flag, and I figured that was the end of this saga. I was mistaken.

A week or so later, I received notice that the deceased's brother had filed a Congressional Inquiry (CI) through Representative Jim Wright's office. I thought it was a pretty lame CI and started to type up a careful response. As I typed, the mail arrived. It included a touching, handwritten letter of appreciation from the widow, insisting her husband was laid to rest with appropriate dignity and honor. I suspect she and her brother-in-law had words over the minorities on the team, and she prudently mailed her letter to thwart an anticipated complaint.

When asked to provide evidence for Colonel Robertson's executive officer to reply to the CI, I included my justification memo and a copy of the widow's letter. Naturally, the major confronted me with snide comments about my non-compliance with what he viewed as a reasonable request. Anticipating this, I had already made him a copy of the widow's letter. I handed it to him and said, "You'll want to read this letter from the widow, sir." He mumbled something unintelligible and waddled away to the latrine. Had we been the same rank, I likely would have followed the major into the latrine and thrashed him for good measure. MSgt Goodwin just shook his head and shrugged, then grinned at me.

Fort Worth had three major breweries, and their boards of directors all had military retirees from Carswell AFB or elsewhere. We received frequent invitations to post and retire the Colors at their significant events and accepted every one that fit our schedule. They always had excellent food and free beer, so my guys jumped at the chance to go. I made certain we had a designated driver (usually me or another NCO) when we worked them. Back then, most military vehicles had common keys and some retirees had kept extra or spare keys. Beyond that, the large lock knobs were easy to reach through the door window gaskets with a coat hanger. At nearly every brewery event, we returned to our van to find a dozen or so cases of beer stashed under the seats. During one late afternoon event at a brewery, we had to leave to work a separate evening event across town, so we could not stay to retire the Colors. When we arrived back at the Services Squadron later that night, I noticed a tarp over a stack of something next to the door. Under it were 20 cases of beer with a "thank you" note from a retired general on the brewery board. I split the cases evenly between my team members that night and then dropped a couple off with my Delta Flight LE buddies on the way home.

We also provided color guards to a lot of high-profile events, including a Dallas Cowboys game and the World Roller Skating Championship. The Cowboys game was cool, and I waited on the home sideline as my team marched across to me. A camera zoomed in on me saluting during the National Anthem, and for a few seconds, I was on national TV. My father and several other relatives and friends saw me and later called to tell me about it.

I started dating a civilian woman in late fall and it didn't last long. Her family didn't like her dating a military man and they treated me with casual indifference — almost as if I were an accessory or not there at all. Her dad was a serious alcoholic and liked to walk around in his pajamas. This woman liked to antagonize, argue to the point of provoking nastiness, and then make up — personality traits I despised. I was certain she had deep racist sentiments as well.

One night at dinner she started mouthing off at me about her angst issue du jour. I dropped her off at her house and said goodbye, adding that our personalities were incompatible and we should not see one another again. I drove off with her standing there yelling at me. She mailed me a nasty and condescending letter, followed by several phone messages pleading with me to call her. I ignored this and never looked back. While I halfway expected her to vandalize my car, she never did, and I was grateful for that tender mercy.

I returned to Delta Flight LE in December, right before Christmas. While I was away with the honor guard, much had changed with the Deltas. SSgt Chaires (K9) had transferred to the Philippines, and several others had transferred or gotten out. SrA Dave Lynch reported in as a new Delta K9 member. Ron Rucker had left months before for SP Investigations, and I didn't see him much. While one K9 guy, A1C Todd Swerske, had reported in the previous spring, he had since transferred to Charlie Flight LE as a replacement. Dave came to us from Korea and was fun to work with; we also enjoyed hanging out, hunting, or fishing while off-duty.

198312: Senior Airman Dave Lynch in Korea prior to arriving at Carswell AFB in December 1983. "Big Dave" was an outstanding K-9 handler. He passed away in Tucson, Arizona in 2020.

199307: Later photo of Senior Master Sergeant Ron Rucker and US Army First Lieutenant Clark at Carswell AFB in July 1993. Dan stopped by to visit Ron while on business in the Dallas-Fort Worth area. Ron retired a few years later and became a Federal employee in Fort Worth. He is also a minister. We stay in touch and reminisce about funny stunts we pulled as junior enlisted men.

During the previous spring and summer, other flights (mainly A Flight LE) wrecked three patrol cars. In December we received replacements and two were police interceptor models. Mine was a Ford LTD II with a 5.0-liter, high-output, fuel-injected engine; it had the full police package (front ram bumper, heavy-duty shocks, backseat cage, etc.) and was very nice. It was far superior to the old beat-up Dodges that we kept running with cannibalized parts. Those were so old that almost anyone could outrun them, especially motorcycles. We typically had to call gates to stop traffic to catch fleeing vehicles—very embarrassing.

Each flight now designated two NCOs to drive for a new traffic detail called Enforcer. I was the primary Delta Enforcer and the other was the flight chief. The speedometer reflected a max speed of 160 MPH. While checking on the NMSA one mid-shift, I asked the Texas DPS (highway patrol) to let me "test it" out on I-20. It accelerated to 100 MPH like a rocket and I backed it off without further testing. There was zero chance a speeder would ever again escape from me on base.

Enforcer duty was created purely for traffic enforcement and accident investigation. Several commanders were upset with speeding on base and felt a flurry of tickets would solve the problem. The real problem was twofold: (1) people not paying attention while driving and (2) pedestrians stepping in front of cars. As children, people are taught to look both ways; yet somehow, around age 12, that turns into "cars have to stop," and people foolishly assume cars will—often to their peril. When a jaywalking pedestrian was hit on Rogner Drive (the main drag) earlier that year, the knee-jerk reaction was to drop the speed limit from 30 MPH to 25 MPH, an illogically slow pace. Never mind that a careless jaywalker caused the accident. As a result, people ignored the lower limits unless a traffic unit was present. Thus, Enforcer duty was born.

Writing a few solid tickets was possible during swing shifts when plenty of cars were on the road. Mids were another story and I got hounded every night to write tickets for something. So on mid-shifts I idled along barracks row and through NCO and officer housing areas to issue tickets for registration violations, cars blocking sidewalks, cracked taillights, cracked windshields, flat tires, or anything else I could find. Although most states allowed flexible extensions for military members whose plates had expired, I was writing 20 tickets per night for this alone. Complaints mounted, and the commanders backed off bugging us about ticket quotas on mids. I still managed to write justifiable tickets for moving violations during swing shifts. After a month or two, the dedicated Enforcer duty went the way of the dinosaur due to more pressing needs and higher priorities.

I had experienced multiple problems (starter, carburetor, etc.) with my Ford LTD. I got tired of it refusing to start and breaking down, so it was time for a new car. I had saved up a lot of money while on active duty and asked my dad for help finding one. A master negotiator, he knew every reputable car dealer in several counties around. He called me in late December to say he had found me a 1983 Ford Thunderbird in Pulaski, and the dealer agreed to $10,000 even. My dad said it was a good deal. I replied, "Please buy it for me and withdraw the money from my bank account. I'll pick the car up next time I'm home on leave."

The SSgt serving as the chow hall's swing shift mess sergeant was arrested for molesting his 14-year-old stepdaughter. He seemed like a decent guy, so his arrest shocked everyone. While in our jail awaiting sentencing, his wife filed for divorce at the Tarrant County courthouse, so several meetings downtown followed. A Flight LE decided to transport "SSgt Molester," his estranged wife, and his wife's lawyer to one meeting in the commander's staff car. The car was an AMC Pacer that LtCol Griffin hated and never drove; thus, it was always available for errands. Sgt Kroger was tagged to drive the motley group to the county courthouse.

A series of mistakes followed that almost turned catastrophic. For some reason, the desk sergeant handcuffed SSgt Molester's hands in front rather than behind his back. With the estranged wife and lawyer on board, Kroger put SSgt Molester up front with him. Kroger also stayed armed instead of turning in his pistol.

No sooner had they rounded the corner from the police desk than Molester grabbed Kroger's pistol from his right-side holster. Kroger let go of the steering wheel to grab the gun. Molester tried to shoot his estranged wife, but the bullet flew out the window. Kroger dodged another round that went by his face. Molester then pulled the gun to his head and shot himself. While the round broke his skull, it was not fatal; the graze wound rattled his brain significantly. SSgt Molester spent the next few weeks downtown in intensive care under SP guard.

A week later, I had to go babysit this clown during a swing shift until the East Gate closed at 1800 and I could get relieved by that guard. Molester was handcuffed to the bed rails. He kept yapping and whining for me to release him. I finally said, "If you don't shut up, I'll use my boot heel and shut your sorry ass up for good." He left me alone after that. A nurse overheard me and got upset, but I didn't care because he was driving me nuts.

A younger Security troop came to relieve me around 1830 since we were still short of LE cops after the Charlie busts. I had just gotten back in my patrol car when the radio crackled that Molester had gotten loose. I looked everywhere for the escapee along adjacent streets, thinking he had fled on foot. Within 20 minutes, the hospital staff reported finding him in a basement locker room. He was transferred to a prison hospital in Dallas until he recovered well enough to return to our base. A day or two after returning to Carswell, he was sent to the U.S. military prison at Fort Leavenworth for a long time.

The young SP who relieved me swore he only let SSgt Molester loose to use the bathroom. However, he could not successfully explain how a fat, wounded guy in a nightgown slipped past him out the door to the stairwell. I included this in my incident report, and the young lad was honest about his mistake. LtCol Griffin let him off easy with a letter of reprimand.

Don Adams left the service in January 1984. He quickly got hired as a trucker and moved back near his parents' place in Irving. He stopped by for a beer occasionally but eventually went silent due to his time on the road. Once Don left, SrA Bruce Crosby became our full-time desk sergeant. I really liked Bruce, and we got along like brothers.

It didn't take me long to find a new roommate, an avionics technician named Charles Aston. We were promoted to Sergeant together, and he was a good roommate. Chuck kept the place clean, so I let him have the master bedroom. He talked me into letting another mechanic move in, whose name I can't remember. He was a huge fellow and would eat a whole beef roast and a family-size bag of French fries in one sitting. We all worked separate shifts, so we were rarely there together at the same time. This was convenient when someone had a date over.

Long before Don left, we had ended the security guard arrangement. When he and Mark left a month apart, I had to start paying their rent shares, and this ended my ability to save money. My new roommates were okay with paying $250 each, so we easily made our $750 per month rent. This was a relief since I didn't want to move alone to an efficiency apartment or back to the barracks on base.

In February, I took a week of leave and drove my old LTD home to Tennessee. In the garage was my new T-Bird. It was solid white with a light blue interior and crushed velvet seats. I fell in love with it and was thrilled to have a car I didn't have to constantly work on. It was comfortable, got good gas mileage, and accelerated like a scalded dog. The car had only 7 miles on the odometer when my father picked it up in Pulaski, so it now showed 27 miles. I had already reimbursed him for the cost and was happy with my investment.

During mid-shifts while checking the marina, I often fed the ducks old bread I brought from the chow hall. One night I decided to grab a couple of ducks for a prank. I stuffed them in the trunk of my patrol car and drove them over to Wing HQ. I put them in the hallway outside the Command Post door and left. I never told anyone other than Bruce Crosby, our desk sergeant. Naturally, the B-52 alert crews got blamed for this harmless prank, and I later found out they even accepted credit for it. As it turned out, the alert crews were notorious for pulling pranks like that.

One Sunday night, a friend over on the Security side of the flightline radioed for me to meet his patrol at a large hangar. When I arrived, he handed me a laundry bag with something wiggling inside. One of his teams had herded a jackrabbit into a B-52 hangar and caught it. They wanted to do "something fun" with it. We thought for a minute and it hit me — night book drop at the base library. I took it there, opened the chute lid, and shoved the rabbit in. The next morning when the librarian opened the book drop door, the jackrabbit sprang out and ran like crazy around the library. It took two responding day-shift patrols to help the librarian run it out the front door. As expected, B-52 alert crews neither admitted nor denied pulling this prank.

One night after I arrived home from a swing shift, my EST beeper went off. I drove back to base and joined my team. A drunk in the off-base housing had beaten up his wife and taken her hostage. We responded to the scene, and by then the hostage negotiator had managed to get the guy to release his wife and kids. She said he was armed with a shotgun.

We didn't want to go in after a barricaded suspect. MSgt Phillips and I discussed the best option, which was to wait him out, reasoning that by morning he would sleep off his drunken state and surrender quietly. LtCol Griffin and Colonel Robertson concurred, so we maintained a perimeter until morning. We were right, and at daylight the man walked out with his hands up.

The Delta LE team started firing on all cylinders that winter and spring. Everyone seemed to get along well together. A new NCO — SGT Jerry Haller — was reassigned to Delta in late winter; he was married with kids and loved baseball. Jerry's only negative was getting tired on mids and hiding to take short naps. One night, I caught him asleep in his patrol car inside a B-52 hangar. His was an older car, so the lights and siren wouldn't come on unless the engine was running. I sneaked up to his car and carefully reached in the passenger window to flip the switches. I got Lynch to bring in his K9 to wake Jerry up. As the dog growled softly, Jerry startled awake and instinctively started the engine. The lights and siren came on, and he got the point. We laughed about this one for days.

Games of "water balloon tag" were also great fun to keep people alert during mid shifts. Usually around 0300, patrolmen would find a spot to turn off their patrol cars and "monitor traffic." This often meant taking a quick nap and those guys were fun to target. If passing a parked patrol car and the occupant didn't blink the lights or do a single click on the radio mike, there was a strong probability he or she was napping. And that's how water balloon tag began. It was so much fun to park nearby, sneak up within throwing range, and splat a windshield to startle the occupant. Yet this could backfire at times, especially if the "target" was setting an ambush. When a new guy tried to tag SrA Lynch, Dave let his K9 out the side door. The poor kid who threw the balloon had to run for his life to get back to his patrol car before the dog tore into him. I witnessed this spectacle and laughed so hard my sides hurt.

A group of us liked to fish during our three-day breaks, notably Bruce Crosby and me. We particularly enjoyed catching white bass (also known as sand bass) that schooled together for dawn and evening feedings. White bass stop growing at around the 2–3 pound range, and most we caught were in that range. They go into a feeding frenzy at dawn and dusk, so we often caught 20 or more each in 30 minutes. One morning we caught over 100 at Eagle Mountain Lake and took the biggest ones home to eat. We cleaned them at Bruce's apartment and dumped the guts and waste into the dumpster. Unfortunately, the dumpster had just been emptied, and the entrails stank like hell before the dumpster truck's next visit. His neighbors were very upset.

One night some unknown perpetrators pulled the drain plugs on several senior officers' boats at the marina. The boats sank at their moorings and were a chore to re-float, haul out, and clean up. The commanders were hopping mad, so the base commander put the marina area off-limits at night, with the exception of the Alert Crew Visitation Center. The marina area and picnic grounds were in the patrol supervisor's normal patrol sector, so while I was glad it didn't occur on my shift, it still annoyed me that someone did it.

It took a while for word to get around that the marina and picnic grounds were off-limits from dusk to dawn. It was a notorious parking area for teenage dependents, so I had to run them away for several weeks. Inevitably, I caught a couple going at it in the back seat of a car right next to the marina. They were officers' kids and very embarrassed to get caught. I didn't report them, just ordered, "Get dressed, get gone, and stay away from the area after dark." Naturally, I became known as "Sergeant Cockstop" and "Cockblock Clark," even among my peers.

Another time I caught a young pilot buck naked with a partially clothed woman in an alert crew truck. Actually, he did have on a Timex and his socks. I let them go too. I knew the single guys lacked the same room privileges at the visitation center as their married peers, who were easy to spot occasionally carting in sheets and refreshments. That month-long alert period was obviously tormenting to the single guys on the alert crews.

During a mid-shift we received an exercise distress call from the Armory. An "armed terrorist" had seized the LE Desk and was holding the desk sergeant hostage. Knowing our radio net was monitored, we had code words for such contingencies. I radioed our patrols: "This is Police 2. Meet me at the O.K. Corral in five. Do not acknowledge." The Wing Commander was Colonel O.K. Lewis, so the "O.K. Corral" was our code word for his HQ. Within minutes, I had three LE patrols on site plus two Security patrols from the flightline. We then switched over to a separate reserved radio frequency.

I posted the 360-perimeter cordon as TSgt Childs arrived to take over our entry control point. I had a Security NCO with a rifle climb up the ladder with me to the SP HQ roof; he would serve as our designated marksman. A skylight above the jail looked directly down the hallway and into part of the desk area. I watched for a few minutes and monitored both the LE and separate frequencies as the negotiations took place. After a little while, we determined the intruder was alone.

Our marksman and I could see through the rooftop windows, and the bad guy moved into the hallway almost under the skylight. I radioed on the LE net, "Look up." When he did, I shined my red-lens flashlight through the window into his eyes. I keyed my radio and said, "You're dead. Game over." At that time, our assault team burst in, grabbed him, and seized control of the LE desk area. Captain Neal was happy with our performance. As Childs lit a cigarette, he joked, "Too bad you don't smoke, or I'd give you one of these."

One midnight while I was checking the North Gate, the radio crackled that someone had run the Main Gate. The car flew past an LE patrol on Rogner Drive and was headed my way toward the lake. I blocked the interior hospital gate near the Officers Club and called the North Gate guard to close the outer gate. Security teams closed all the flightline gates and within one minute all gates were shut. The gate runner was trapped on base and wasn't going to get away.

SrA Lynch intercepted the runner south of the lake and a wild chase ensued through base housing. The runner hit a dead end and cut across several lawns, smashing an above-ground swimming pool as kids' toys flew into the air. He circled around, got back on the road to the lake, and sped in my direction. I had my pistol drawn and was cleared to use deadly force to stop the vehicle. Then, for some reason, the runner stopped at the lake road's T-intersection, possibly out of confusion due to the dark lake ahead. The car turned right in my direction, so I jumped into my car and intercepted the runner from the front with lights flashing. I stopped a foot from the intruder's bumper to prevent the car from fleeing, so he was now boxed in and surrounded. I got out and took cover behind my door.

Lynch came up quickly with his K9 and ordered the occupant to turn the car off and throw the keys out the window. The driver complied and sat silently. I ran up and we yanked a large bearded man—reeking of alcohol—out the door and handcuffed him. Captain Neal drove up and told me to read him his rights and place him under apprehension. The man didn't like this and started yelling curses at us. When I asked if he understood his rights, he retorted, "Do you?" We ignored his cursing and I stuffed him into my patrol car. The only thing he said en route to the LE desk was, "Hey, Sarge, go easy on me. I served in Vietnam." I didn't respond, figuring he was too wasted to understand anything I said.

We booked him and during interrogation determined he was just very drunk and had taken the wrong turn off Highway 380. When he came to the Main Gate, he got confused, then panicked and hit the gas. There appeared to be no ill intent on his part, and he wasn't armed. When I went back to search the car, all I found was an old pocketknife under his driver's seat. It wasn't in easy reach, so I wiped my prints off it and pitched it into his trunk.

At daybreak the next morning, Fort Worth Police came to retrieve him. By then the man had sobered up considerably. On the way out the door he turned to me and said, "I'm sorry for the trouble I caused. I'll be glad to pay for any damages." While I felt sort of bad for the guy, he had no business driving drunk like that.

Per the latest strategic arms treaty, Soviet satellites passed over our SAC bases twice daily to count the number of B-52s in the alert areas. A satellite passed over Carswell AFB every 12 hours at approximately 1105 and 2305. So at 2250 each night the alert area lights came on and stayed on until 2320. This gave the Russian satellite enough time to confirm how many B-52s were in the alert area. If it was overcast, Soviet spies could count the B-52s from private aircraft or from the bluffs across Lake Worth. There was a white X painted on the pavement of the base armory parking lot; if you stood on it and stared up at the tip of the armory's radio antenna on a clear night, you could see a faint pinpoint of light slowly pass by high above. That was "our" Soviet satellite.

198403: Sergeant Clark briefs a patrol at the Carswell AFB armory in March 1984. The annoyed look was somewhat standard for everyone.

And this was also the source of several incidents when the lights came on in the alert area. When Soviet leader Leonid Brezhnev passed away in late 1982, someone unfolded a poster that read "RIP, Leonid" on top of a B-52 wing. When his successor Yuri Andropov died in 1984, someone else draped a bedsheet that read "See you in hell, Yuri" on another bomber. And of course, occasionally some SP would think it funny to moon or "wiener-wag" the sky as the satellite passed. Security grunts in the alert area also liked to make funny faces while crotch-grabbing and faking lewd acts in hopes of becoming "that guy" who gained notoriety for insulting the Soviets. While it was a little harder to prove culpability for the signs, it was very easy to figure out the others who were clowning around. Few of our guys realized how powerful satellite imagery was in the early 1980s. Both the U.S. and Russia had satellites that could take space-based photos with such clarity as to tell the time on a gate guard's watch.

There was another mindless game called "wing walking" that earned offenders an automatic Article 15 when caught. These clowns would jump up on the tip of one wing, walk across the bomber, and jump off the opposing wingtip. While this nonsense was unlikely to damage the aircraft, the B-52s in the alert area were loaded with nuclear weapons. It was just a dumb thing to do and severely damaged one's reputation and career.

So what? Here's what: if you did something stupid while Soviet satellites were overhead or spies were watching, it was a good bet the Soviet Ministry of Defense would have the Soviet State Department send the photos to the U.S. State Department, which would in turn forward them in descending order to the Department of Defense, HQ U.S. Air Force, HQ Strategic Air Command, 8th Air Force, the 7th Bombardment Wing, and 7th Security Police Squadron. The wing commander chewed out LtCol Griffin, who in turn chewed out the shift commander, who then ripped into the offending airmen and their supervisors. The takeaway was to keep your private parts zipped up tightly and maintain a degree of professional dignity, especially during the time windows those Soviet satellites were passing overhead.

OK, with that said, I'm sure Soviet troops were doing the same thing as U.S. satellites passed over their fixed missile silos and bomber bases. I suspect the difference was the Soviet offenders likely received merit promotions for insulting America. Likewise, I'm quite certain our satellites were equipped with telescopes that strained mightily for our intelligence and imagery analysts to even see the commies' pitiful little meat puppets—except on their women, whom we suspected made up for their male counterparts.

In addition to the scheduled exercises, SAC bases conducted frequent no-notice alerts. These caused the alert crews to scramble to their B-52 bombers and KC-135 refuelers in the alert areas. The LE patrols sped to their designated flightline gates and allowed only the alert crew vehicles to pass. The 7th Bomb Wing conducted these exercises so often that our required response actions became routine; unless announced as exercises once initiated, we were never sure if they were live missions. Once the bomber crews were on board, they fired up the engines, and the command post typically declared

"ENDEX." At that point, everyone stood down and resumed normal operations. However, occasionally the bombers conducted an "elephant walk," where they left the alert area, taxied out to the north hammerhead, went down the runway to a cross-taxiway, and then reentered the alert area. We saw a few of these, too.

198210: This is a B-52-D model bomber at Carswell AFB, Texas in October 1982

198210: This is a B-52-H model bomber at Carswell AFB, Texas in October 1982

Then came the day that the exercise appeared to be the real thing. An alert kicked off, and I drove to my flightline gate to let in the alert crews. It was late afternoon, so I had to stop and hold a lot of cars in place, mostly dependents and retirees en route to or from the BX and commissary. The alert crews hauled tail into the alert area and in minutes the B-52s were "walking" out to the north hammerhead. Only this time they halted in place momentarily as if awaiting orders. We knew if the bombers took off carrying live nukes, World War III was underway. That concurrently meant Soviet nuclear missiles were inbound, and we had 25–30 minutes before they struck and vaporized us.

The lead bomber fired up its engines, and black smoke billowed from it as it started to take off, followed by the second bomber firing up its engines. I looked at the faces inside the cars lined up in both directions and saw people visibly upset, a few crying. They knew what I did — that this might be the big one.

No sooner had the lead bomber hurtled out of the hammerhead to take off than it got partway down the runway and cut back its engines to slow down. The rest of the alert bombers followed suit, gunning their engines to trot down the runway before lumbering back up the parallel and cross taxiways to the alert area. The LE desk radioed for us to release traffic, and I could see the sense of relief on the faces of drivers and passengers as they passed by. Once the traffic cleared, I stepped over to the curb and vomited violently for a couple of minutes. That's a fair description of the stress level we felt serving on a SAC base during the Cold War. It was a serious mission with imminently deadly consequences if WWIII ever began.

A Delta Security guy named Winston "Win" Woodall was on my EST team. He was an interesting character who survived the urban ganglands growing up. He escaped that lifestyle by enlisting. He had numerous scars, including one rumored to be from a gunshot wound, and was also rumored to have killed several rival gang enemies. Win didn't talk about his past and we didn't ask—hence all the rumors. A heavily muscled Black kid (5'10"/ 230), he had a Monster Bench in his barracks room and worked out constantly. He was impressively intellectual, into photography, and read everything he got his hands on, including the classics and poetry. Win's favorite weapon was the shotgun and he was the rear overwatch man on our team. If he was afraid of anything, we never figured out what it was. Although he never talked much, Woodall struck me as a competent SP and was pleasant to have around.

That said, Win had a penchant for getting into minor trouble—usually on others' dares. During a Commander's Call at the base theater during a 24-hour break, he changed the movie listings on a menu board to off-color titles by rearranging the letters, made easier because the theater manager had left the letter box on the counter.

Black Beauty became Black Snatch; Revenge of the Ninja turned into Revenge of the Nasty Anal Pussy. Close Encounters of the Third Kind became Denture Job – Anal Encounter. While these are the only ones I remember now, I noticed a lot of guys standing in the lobby laughing about something. I didn't see the menu board following Commander's Call, but I saw the evidence photos later. The theater manager discovered the prank after he opened for business that evening and blew a gasket. He called Colonel Robertson, who called LtCol Griffin, who called Captain Neal, and so on down the line. Bad dookie tends to gain velocity exponentially as it rolls downhill.

That night at our midnight shift guardmount, Captain Neal had an out-of-body experience over it. Despite most having no clue about the incident, a few snickers from within the combined formation indicated there were many witnesses. When no one stepped forth to take responsibility, Captain Neal warned that severe consequences would follow for the guilty parties. After guardmount broke, Woodall went to see Captain Neal and confessed his guilt. He accepted an Article 15 as punishment. Two NCOs who egged him on received severe reprimands as well.

It didn't take Win long to get in trouble again. He arrived home from EST training one weekend night to find his civilian roommate had thrown a party at their apartment. On entering his room, Win discovered his girlfriend on his bed with another man. He pulled a rifle (Mini-14, I think) from his closet and ordered everyone out. In minutes, he heard sirens and took off on foot into the adjacent neighborhood.

I had just gotten to sleep when my EST beeper went off. I headed out to help search for Win, and several of us searched for a while before calling it off. He was street-smart and knew how to hide. The next morning we were relieved to learn Fort Worth Police picked him up as he was walking home with his rifle. He had climbed into a big pickup truck's bed and fallen asleep, only to wake up the next morning when the truck started moving. Reluctant to jump from a moving vehicle, he didn't alert the driver and stayed hunkered down in the truck bed. Win rode for a while and finally jumped out at a stoplight to begin the desperate run home. Already with an "all-points bulletin" in effect, it's not hard to imagine how the local police departments responded when they heard, "Large Black man with rifle running west along XYZ Street."

Woodall was up for promotion and had transfer orders to Germany. He was a solid SP and I didn't want to see Win get kicked out, especially since he was goaded into this incident by a cheating girlfriend. I floated an idea by Captain Neal, who agreed to let me present it to LtCol Griffin. I suggested Griffin assign Woodall to me, and I would put him to work at the K9 kennel. I reasoned the probability was low of Win getting in trouble while feeding the dogs and hosing poop from the cages. Beyond that, we needed a new kennel attendant on Delta anyway. To my surprise, LtCol Griffin agreed and gave SrA Woodall to me and the other Delta LE NCOs. However, he warned me, "If he screws up again, it's on you, Sergeant Clark." Win promised good behavior and he was true to his word. He transferred to Germany a couple months later.

In mid-spring, I received my letter of acceptance from Middle Tennessee State University for fall 1984 enrollment. Whereas my parents wanted me to get out and attend college, everyone in my chain of command urged me to reenlist because they were certain my name would appear on the SSgt list about to be released. I had trained for two years to attend the U.S. Army Ranger School and had just received a tentative class seat for late October 1984. However, reenlistment was required for me to go. I was torn between getting out to attend college and going to Ranger School, since it was a prestigious career opportunity for an Air Force SP — or anyone else, for that matter.

I was still mulling over reenlistment and was a week away from the decision deadline when I received assignment orders to Ramstein Air Base, West Germany, for October 1984. These orders were also contingent on me reenlisting. Libyans and European communists were attacking Americans and blowing up discos frequented by U.S. troops all over Europe, so I was told the Germany orders took precedence over Ranger School. When the base personnel office told me Ranger School was definitely out and I had to accept the Germany orders, I declined to reenlist and signed my intent to leave active duty. To start college on time, they approved a 30-day early release so I could enroll in August. Minus my leave already approved for June, I'd have 72 days of terminal leave, and my release date was set for July 13 — Friday the 13th.

SrA Lynch called in one night that he discovered a blonde woman with a British accent covered in blood near the NCO Club. We transported her to the base hospital, where we learned she was married to one of our Delta Security NCOs — SSgt Watson — who was on leave. We couldn't believe it since everyone considered Watson a great NCO. It took a while for the doctors to get her husband's name out of her; she admitted he got drunk and beat her up. By then I was at the hospital and heard her wail, "I left a great 30,000-pound-per-year job in England for him and now he does this to me!" She confirmed Watson was at their home. We knew he had guns and this might get dangerous.

Watson lived in base housing near the NCO Club. Five of us, including Captain Neal, went to find him. As a big-game hunter and survivalist type, we suspected he kept at least one loaded firearm. His wife warned us that he kept a loaded pistol on the bedroom nightstand. Using his wife's keys, we quietly entered the home. There were blood drops all over the kitchen, hallway, and bathroom where his wife had tried to escape. Bloody smears marked the wall where he ripped the phone out as she tried to call for help.

We were annoyed but calm as we entered the bedroom. Watson slept motionlessly as four pistols pointed at him and Captain Neal flipped on the lights. He stirred, blinked a couple of times, and asked, "What the hell are you guys doing in my house?"

The captain asked him to get his hands where we could see them. Watson delayed at first until MSgt Smith cocked the hammer on his pistol and said, "Please do it now, Watson." This gained compliance, so we grabbed and cuffed him. He was buck naked except for a wristwatch, so we pulled shorts onto him and took him to the desk. During our search of his quarters, we found and confiscated his pistol and several other firearms for safekeeping. He was in a heap of trouble, and that bothered me greatly because I really liked Watson. Always a solid performer, he treated everyone with respect and trained the younger guys well.

During the interrogation, Watson (obviously still under the effects of alcohol) mouthed off a lot to Captain Neal, which made his situation even worse. His Security shift chief, MSgt Bobby Joe Vandegriff, had come to talk to Watson and witnessed everything. However, the damage was done and the chief finally walked out shaking his head in disgust.

Stripped of his SP credentials, Watson was placed on administrative duties and put to work at the chow hall while awaiting his hearing. He realized he goofed, and to his credit, he never tried to blame anyone else for it. Not considered a flight risk and now under supervision of the base's attorneys, mental health team, and chaplain, Watson was allowed to remain in base housing at his wife's insistence. Within 90 days, he was discharged. We were told his wife decided to stay with him and last we heard they moved back to England. Except for this lone unthinkable incident, Watson was a good NCO and I hated to see him go.

Early in the spring, Captain Neal informed me I would represent the 7th SPS as its nominee for "8th Air Force Outstanding Security Police Airman of the Year" for 1983. He was also representing our unit as the Security Police Company Grade Officer of the Year. The 8th Air Force board selected both of us as its recipients and forwarded our packets to the Strategic Air Command board. Because I lacked off-duty community service activities, I came in second place at the SAC board, but Captain Neal finished first. He was ultimately selected as the Company Grade Officer of the Year for both Security Police and the U.S. Air Force overall.

198401: Captain Donald W. Neal, Jr at Carswell AFB, Texas in early 1984. He was among the most omni-competent officers under whom I served over three decades. He was named "Company Grade Officer of the Year, 1983" for the entire US Air Force. Don retired from the Air Force Reserve as a full colonel and has served as an attorney for the State of Texas for many years.

We had numerous interviews and received significant media coverage. Our photos appeared in newspapers and Air Force publications. While I received a nice plaque, Captain Neal was given all kinds of accolades since his awards were so significant. One of the breweries threw a big keg party for him; he didn't drink, so instead invited all of Delta Flight to it.

198406: Delta Flight Law Enforcement guys at the armory waiting to go on duty in June 1984. Sitting is SrA Bruce Crosby (desk sergeant). Standing (L to R) is A1C Bryant, Sgt Jerry Haller (almost obscured behind Bryant), A1C Greg Sly, A1C Sherman (Delta Flight Security), A1C Ross Wood, and A1C Josie Dempsey. Many left active duty to pursue careers in civilian law enforcement. Bruce made it a career and retired as a Master Sergeant. While I occasionally hear from Sly, Bruce and I communicate at least weekly to this day.

Because one of my emerging wisdom teeth was impacted, I had to have them all removed. The surgery was at 0800 one morning after a midnight shift. I didn't want to go under anesthesia, so the dental surgeon (a full colonel) agreed to let me stay awake for the procedure. No sooner than the second tooth was out, the klaxon went off for the expected initiation of "Global Shield," the annual worldwide nuclear readiness exercise. The surgeon ordered, "Stay put, Sarge. I have to go sign in." I thought, Geez, doc—just finish the damned procedure!

I lay there almost upside down for several minutes while blood trickled down my throat. Swallowing it nauseated me, and I had to throw up; I made the mistake of eating breakfast at 0700, so it came up too. The surgeon walked back in and was instantly angry over me barfing on his operating room floor. He took out the other two teeth, handed me some pain pills, and yelled at me to get the hell out.

I was still loopy and bleeding from the surgery, so I signed in at the Armory before going home. I was told to return for duty at 1800, so I went home and fell asleep on the couch. I signed back in at 1800 but was in no shape to stand guard on the flightline. Captain Neal had me posted in the WSA instead. While assigned to radio duty, I kept spitting up blood, and my mouth hurt.

The NCOIC told me to go sleep in a closet. I was really tired and didn't mind crashing out on the A-bags full of combat gear. Around 2330, Captain Neal visited and told me to go home for the rest of the exercise. He said I was to stay home for the next two days, and he would cover for my absence. I was grateful to him for it, and the time off helped me recover faster.

In February, I started dating a green-eyed redhead named Shay. She worked in the OB/GYN clinic at the hospital with Lisa Richards, who was (I think) her immediate supervisor. Lisa had introduced us a few months earlier at a dance club, and we seemed to click. Afterwards, I pursued Shay for several weeks in vain before giving up.

Not long afterward, during a swing shift, I was driving past the base hospital and spotted her leaving the building. I stopped and offered her a ride to her barracks. She declined because she already had a ride, then surprisingly asked me out.

Shay was of full-blooded Irish ancestry and four years my senior. We had solid chemistry, liked the same foods and music, and enjoyed one another's company. The only problem was that she worked a straight Monday–Friday day schedule, and I worked rotating evenings and nights. Since we were rarely off at the same time, getting together for more than a few hours at a stretch was a challenge. What nagged me was the realization that things might abruptly end when I got out in July. Ironically, she had requested a transfer to Ramstein AB, Germany, and I had just declined that very assignment so I could get out to attend college. That didn't bother us, and we made the best of it.

Bruce Crosby and his wife, Lisa, introduced me to her cousin, Cheryl, in mid-April. She had been in several abusive relationships and they wanted her to meet someone nice. She had recently moved back in with her parents and was working in an entry-level retail job. Her parents screened her calls to prevent a persistent ex-boyfriend from contacting her. Lisa insisted I was a better fit for her cousin Cheryl than Shay, so I made a somewhat impulsive decision. I let Shay go and she completely broke contact with me—not that I blamed her.

All my hospital friends, except for Lisa Richards and Betsy Fowler, were also very annoyed with me for breaking it off with Shay.

Cheryl and I hit it off well and started dating. Her dad was retired military and worked on base, so she could visit me at work anytime she wanted. In May, she got a black eye playing softball on her league team. Most assumed a boyfriend had done it—something she felt compelled to explain to everyone we met. That should have served as a red flag, but it didn't register as one at the time. I was a gentleman toward her and everyone thought we made a great couple. Her parents and I grew close and things seemed very right for a few weeks. We decided she would go to Tennessee with me on leave for a few days in June.

One evening in early June, I arrived at her home to find Cheryl arguing in the driveway with a very large blond guy in a dual-wheel pickup truck. We were supposed to have dinner before my mid-shift, and I was in uniform. I presumed he was her most recent ex-boyfriend, who was allegedly physically abusive. He was still bellowing at her as she ran into the house.

He then turned on me and volunteered that she was still calling him, and that she had recently dropped off a photo at his home in the next county. Knowing she had just had these photos made, he had no reason to lie about it. She yelled at him from inside to leave. He yelled back obscenities at both of us, glared at me, and got in his truck to drive away. I stood there in shock, realizing Cheryl had lied to me about completely severing contact with this jackass.

Suddenly, the guy stopped and exited his truck. He pulled a 2x4 from his truck bed and started across the yard toward me, threatening bodily harm. As he closed to just a few feet away and raised the board over his head to strike, Cheryl's dad pulled up in his own truck at that moment and ordered the man off his property. The guy complied and as he left, yelled, "Get out while you can!" He burned rubber as he drove away.

I apologized profusely to Cheryl's dad over this drama. Despite assuring me he wasn't upset with me, he felt it was best if I left. I later learned he had threatened to kick Cheryl out over lying to her parents and me. I should have broken it off right then, but she begged me to stay with her. Whereas Lisa also asked me to give her cousin another chance, Bruce suggested I end it. So I ignored my better judgment and stayed on. I still had a nagging sense of doubt in my mind about Cheryl, especially since I had broken up with someone else to go out with her. I decided to let things ride and see where they went.

I took Cheryl home to Tennessee with me on leave in mid-June. I had been accepted to enroll at MTSU for the fall semester, so we went to Murfreesboro and visited the campus. Along with her parents, we discussed the possibility of her relocating to Tennessee for a while to escape her emotional past in Texas. My dad offered to find her a job at his company headquarters in nearby Lavergne. Overall, we had a good time visiting my family and friends. Things were fine, and Cheryl seemed completely relaxed—until we left to drive back to Texas. She was very quiet and acted like a stranger the whole way, then started crying when we crossed the Texas state line. While I sensed this was a bad omen, I said nothing.

With just weeks left before I started terminal leave, Cheryl grew increasingly aloof toward me. One day, while at the hospital completing my discharge physical, I saw Shay pass by the waiting area. When I said hello, she turned and said, "Your blonde girlfriend is a real piece of work. Serves you right she's going to dump you." That really stung, and I was annoyed with myself as Shay flipped her red mane and walked away laughing. It dawned on me that Cheryl was still using the base hospital on her father's retiree dependents plan. It was obvious they had met and talked.

During my final week on Delta Flight before out-processing, I was picked, along with SrA Allen from my EST team, to teach an Air Force Reserve SP unit how to use MILES gear. During the demonstration, the two of us wiped out half the reserve unit members with our laser gear before they figured out how to maneuver against us. It was a solid training event.

Later that week, Captain Neal finally "killed" me in a flight training exercise. After climbing atop the armory through the roof hatch, he threw a shaken can of warm Coke at me. It exploded on the pavement a yard in front of me as I maneuvered between equipment trailers. The Coke splattered all over me and I was dripping wet with the sticky liquid. My captain had never managed to bag me in two years, so he was quite pleased with himself. We both knew I got nailed due to complacency, since I was days away from leaving active duty. I let my guard down and paid the price in ribbing from my "soon-to-be-former" peers.

Neal and I stayed in touch and he was a key influence in my decision to take ROTC in college. Yet in all the years since, he has never let me forget the "Coca-Cola ambush" on that warm night in 1984. He left active duty to attend law school the following year and served as the general counsel for several organizations in the Texas government. He went on to become a counter-terrorism expert and eventually retired as a full colonel from the USAF Reserve.

A former 3rd SPG colleague showed up at Carswell AFB during my final week there: "Mo" Lester Walker. He was a very light-skinned Black SP who worked in SPG Training, always a good guy in my opinion. He was promoted to SrA shortly before I arrived in the Philippines and then promoted to Sgt a year later. While working in 3rd SPG Training, he had time to study professional development courses—plus he was teaching them to new guys—so he scored 100% on all the exams. He received a highly competitive "below the zone" promotion to SSgt right as I transferred to Texas in 1982. I recall he was married to an Air Force captain, yet that had nothing to do with his rapid rise in rank.

Anyway, as I was out-processing, I encountered him and discovered he was now a TSgt after yet another below-the-zone promotion. His was the fastest rise from E4 to E6 I ever saw while in the U.S. Air Force. After retirement, he went to work for the Bureau of Engraving and Printing in Fort Worth, as did Ronald Rucker after he retired.

I out-processed the week after July 4th. An E4 who was on the base honor guard moved into my apartment slot and even took over the lease on the rented furniture. I moved out one day and he moved in the next, so Bruce Crosby let me sleep on his couch the last couple of nights I was in Texas. While it was hard telling everyone goodbye, my mind was focused on getting home safely to Tennessee and starting college. Captain Neal invited me to stop by the armory and say goodbye to the Deltas during their swing-shift guardmount. Predictably—and to the chagrin of Captain Neal and the senior NCOs—my buddies used this opportunity to yell insults of the "go home and [fornicate] your cows" variety that made me laugh.

The evening before my final clearing with the base personnel office, Cheryl and I had dinner. She barely spoke and I knew things were over between us. As expected, she broke up with me in the car two miles from her parents' home. Upon arriving, and without looking at her, I calmly asked her to please get out. Instead, she tried to explain why this was necessary. I turned to her and again said, "Please get out of my car." She continued to try to talk. This time I yelled, "Get out and go back to your exes!" Cheryl got out and stood in the driveway as I drove away. I felt like I could not leave Texas fast enough at this point.

The next morning I reported to the base personnel office and signed my final paperwork. I was happy to have my DD-214 in hand stating "Honorable Discharge." I arranged to join the Tennessee Air National Guard for the remaining two years on my contract. The personnel NCO gave me my records in a sealed envelope and thanked me for my service. That was it—I was done. I packed my car with what it would hold and spent my second night on Bruce's couch.

Before daybreak the next morning, I left for Tennessee and arrived home at our farm before dark. It was Friday the 13th—perhaps not such an unlucky day after all.

I settled back into civilian life without regret. I worked on our farm, hunted, and fished to pass the time until I moved to Murfreesboro. Because I couldn't fit everything in my car when I left Texas, I left half a carload with Bruce Crosby to pick up during a later trip. When I drove back in August to retrieve my things, I stopped by to see Cheryl at her job. She clearly didn't want me there, so I left and went to dinner with Bruce and Lisa. I was driving east before daylight the next morning and thanked God I was free.

I returned home and worked on the farm until it was time to start college. It was good to be home again with family and friends, and I was eager to get started on my next adventure in life.

About the Author

Dan Clark enlisted in the US Air Force in 1980, got out for college in 1984, and later returned to active duty as a US Army officer. The initial four-year odyssey turned into a 31-year career, including multiple overseas adventures and two tours as an ROTC instructor. He then taught public school for a decade before a difficult cancer ordeal forced him to retire. Now fully retired with grown children, Dan enjoys hunting, fishing, wood crafting, carpentry, and simple living with his wife and their dogs.

www.ingramcontent.com/pod-product-compliance
Lightning Source LLC
Chambersburg PA
CBHW050248010526
44107CB00003B/241